T0122211

Lecture Notes in Networks and Systems 674

The series "Lecture Notes in Networks and Systems" publishes the latest developments in Networks and Systems—quickly, informally and with high quality. Original research reported in proceedings and post-proceedings represents the core of LNNS.

Volumes published in LNNS embrace all aspects and subfields of, as well as new challenges in, Networks and Systems.

The series contains proceedings and edited volumes in systems and networks, spanning the areas of Cyber-Physical Systems, Autonomous Systems, Sensor Networks, Control Systems, Energy Systems, Automotive Systems, Biological Systems, Vehicular Networking and Connected Vehicles, Aerospace Systems, Automation, Manufacturing, Smart Grids, Nonlinear Systems, Power Systems, Robotics, Social Systems, Economic Systems and other. Of particular value to both the contributors and the readership are the short publication timeframe and the world-wide distribution and exposure which enable both a wide and rapid dissemination of research output.

The series covers the theory, applications, and perspectives on the state of the art and future developments relevant to systems and networks, decision making, control, complex processes and related areas, as embedded in the fields of interdisciplinary and applied sciences, engineering, computer science, physics, economics, social, and life sciences, as well as the paradigms and methodologies behind them.

Indexed by SCOPUS, INSPEC, WTI Frankfurt eG, zbMATH, SCImago.

All books published in the series are submitted for consideration in Web of Science.

For proposals from Asia please contact Aninda Bose (aninda.bose@springer.com).

Herwig Unger · Marcel Schaible
Editors

Real-time and Autonomous Systems 2022

Automation in Everyday Life

 Springer

Editors
Herwig Unger
Lehrgebiet Kommunikationsnetze
FernUniversität in Hagen
Hagen, Germany

Marcel Schaible
Lehrgebiet Kommunikationsnetze
FernUniversität in Hagen
Hagen, Germany

ISSN 2367-3370   ISSN 2367-3389 (electronic)
Lecture Notes in Networks and Systems
ISBN 978-3-031-32699-8   ISBN 978-3-031-32700-1 (eBook)
https://doi.org/10.1007/978-3-031-32700-1

This Springer imprint is published by the registered company Springer Nature Switzerland AG
The registered company address is: Gewerbestrasse 11, 6330 Cham, Switzerland

Preface

The present volume is unique in several ways and, in a certain sense, also represents a new beginning. For the first time, readers will find the contributions to the conference 'Echtzeit 2023 (Real-Time Systems 2023)' and selected papers from the international conference 'Autonomous Systems' in one volume, and also for the first time, the contributions to the German-language conference 'Echtzeit' are published in English.

Both conferences, organized by the Real-Time Systems Group of the German Informatics Society, were held pandemic exclusively in attendance in 2023 for the first time since the COVID-19 pandemic period. In both cases, the organizers fight for old and new audiences and authors, in both cases they try to help new scientific findings to spread in the scientific community and last but not least wanted to support the building and maintenance of networks of scientists through a variety of discussions and new contact opportunities. These concerns are even more difficult to realize in the post-pandemic era, in particular through the permanent presence of online meeting systems and newly built electronic community systems.

It is only logical that the partly joint organizers of both conferences try to achieve this task by being more effective and by joining forces. The present volume realizes the demand of the younger generation of scientists to present results in real-time research in English and thus reach a wider audience and, with the transition to Springer as publisher, a more international audience is created for the contributions to the topic of autonomous systems. For both conferences, the now joint conference proceedings means both a reduction in work and costs and the possibility of getting more papers in hand to read without more effort.

The first part of the present volume comprises all contributions to the conference 'Real-Time Systems 2023', traditionally held in Boppard on the Rhine. The series opens with Peter Holleczek's contribution to the 'Real-Time Archive', which he has been building for many years as one of the first participants with a huge number of experiences in relevant research. His unique review of developments, especially of the programming language PEARL, was an excellent opening of our conference. The winner of the student competition 2023, Mr Jan Knoblauch, shows in his paper on deadlock detection in PEARL that the research direction pursued since the 1970s is still relevant today and can inspire young researchers.

In the following contributions, authors have their say on many aspects of real-time systems and hardware, especially in industrial systems. The number of contributions submitted under the topic of safety and security proves the increasingly important role of security aspects, especially for real-time systems in the daily use, which are—in particular—also considered, when designing software with PEARL. Presentations on current applications round off the conference program. The second part of the volume contains selected papers from the international conference and doctoral seminar 'Autonomous Systems', which has been held by the working group 'Real-Time Communication' in

Mallorca at the end of October almost each year since 2008. Although the number of participants at the conference is much larger, the number of contributions is much smaller, since many participants of the conference come to Mallorca to make new contacts and discuss new ideas in detail, which also inspires young scientists and doctoral students. In addition to Janusz Kacprzyk, Zhong Li and Peter Kropf, Stefan Pareigis was one of the keynote speakers who was present at both conferences and whose therefore extensive contribution represents a combination of both events. With his topic 'Autonomous Driving', Stefan Pareigis also shows new and common challenges for both real-time and autonomous systems at the same time. An interesting combination of contributions of topics such as machine learning, fault tolerance and decentralized systems rounds off the presentations, in which, of course, innovative ideas on managing the flood of information in the WWW of tomorrow must not be missing.

The editors of this book 'Advances on Real-Time and Autonomous Systems' hope that you will enjoy reading this book that you might be able to experience a bit of the inspiring moments of both conferences and, finally, that you may get a few own ideas by reading it.

Of course, we would be pleased to welcome one or the other of you at one of our conferences in the next year, maybe also as an author of a contribution. For the conference 'Autonomous Systems', which will take place from October 22 until 27, again in Cala Millor in Spain, all information is available at https://www.confautsys.org, while the 'Real-Time Systems' event traditionally takes place again on November 15 and 16 in Boppard on the Rhine and everything important can be found at the web address https://real-time.de/tagungen.

January 2023

Herwig Unger
Marcel Schaible

Contents

AutSys

Real-Time

The Real-Time Systems Archive

Peter Holleczek[(✉)]

Friedrich-Alexander University of Erlangen-Nuremberg, National High Performance
Computing Centre (NHR@FAU), Martensstr. 1, 91958 Erlangen, Germany
`peter.holleczek@fau.de`

Abstract. Having survived the end of the "paper" era, historically quite
arbitrary documents from the inventory of the real-time systems area's
founding fathers are currently digitised and included in a subarchive of
the general archive `dl.gi.de` maintained by the German Computer Society (GI). This real-time systems archive is structured like a tree. Deep
down, it branches out into subareas, each of which contains an explanatory summary. Contributions usually include their full text, an abstract
and a citation. Proceedings are represented both entirely and by individual contributions. In its current form, the archive shows the enthusiasm
of the developers, and also economically successful implementations of
real-time topics in the 1970s and 1980s. Obviously, this GI subarchive
has arrived on the scene as search engines now prefer it when searching
for real-time topics. Its content and how to deal with it are presented
below.

Keywords: Real-time systems · real-time programming languages ·
process control computing

1 Introduction

Real-time systems, like IT as a whole, were born in the "analogue" world. While
it is said that the IT product internet does not forget anything, the analogue documents of the early days are threatened with the "analogue end". The founders
of the real-time community are retired by now or have left us completely. Their
bookcases have been cleared. Content was preserved more by chance, the compilation of which was rather random. As most of the basic considerations of that
time are still valid today, however, it would be good to have an archive saving
what can be saved.

2 Real-Time Systems Within the GI Archive

In Germany, the real-time community is organised in the specialist committee
"Fachausschuss Echtzeitsysteme" within Gesellschaft für Informatik (GI). The
core of this community can be traced back to the development of the real-time
programming language PEARL in the 1970s, but has, long since, opened up

H. Unger and M. Schaible (Eds.): Real-Time 2022, LNNS 674, pp. 3–12, 2023.
https://doi.org/10.1007/978-3-031-32700-1_1

to real-time systems in general. Fortunately, GI had recognised the problem of "forgetting" knowledge related to computer science and, as a remedy, set up a digital archive (https://dl.gi.de), which is open to all specialist groups, thus following the example of ACM (https://dl.acm.org).

The GI archive is particularly dedicated to "grey" literature, which, apart from scientific publications, opens up content important for practitioners. The portal has a German language interface, only. Naturally, it contains many documents written in German, especially those from the development phase. As the GI archive covers many topics, it has a tree structure leading over several levels to the particular areas such as the one on real-time systems. So far, the contributions archived there are largely limited to historical documents up to the end of the 1990s. This content is to be understood from its historical context.

3   Historical Background

The real-time systems archive essentially breathes the zeitgeist of the 1970s and 1980s. Computers could be used for the first time as control tools for technical processes, both in the industrial and scientific world. Completely new dimensions opened up for users, even if the computers were room-sized in the beginning. Compared to discrete circuits, for example, their flexibility was essential. The process control computers of that time were initially adaptations of classic computers for sequential tasks. They were extended to include options to react to external events (interrupts) and to control actuating variables. What was lacking was suitable programming and associated programming systems. Soon it became clear that assembly language programming was out of the question for complex contexts. A new way of thinking in parallel cyclic processes, synchronisation, selected error handling, and interrupt reactions was appropriate. It was the heyday of international real-time language development.

In Germany, the Federal Government recognised the need for and the opportunities of generous public funding, manifested in the Federal Government's data processing programmes [6] at the beginning of the 1970s. Along with this came the call for standardisation, both nationally and internationally. Funding beneficiaries were the rising and globally active electrical industry, industrial outfitters, computer manufacturers as well as the scientific world with its emerging information technology and computer science scenes. One target of funding at this time was the implementation of the *P*rocess and *E*xperiment *A*utomation *R*ealtime *L*anguage (PEARL).

4   Description of the Archive

Currently, the real-time systems archive contains some 600 items.

4.1   Inventory

The starting stock was mainly fed from the more or less preserved paper collections of the start-up scene. PEARL inevitably appears in various places, and can be used here as an example. The inventory includes everything that was used at the time for language development, implementation and application, constantly with an eye on developments elsewhere. It extends in time to the entry into the digital age with desktop publishing and internet presence at the end of the 1990s. Stricter publication guidelines make it difficult to place them all in an open full-text archive.

4.2   Digitisation

Wherever possible, the documents were disassembled and scanned page by page, with a resolution of usually 600 dpi. Text recognition was turned on provided the font was reliably recognised and the layout was not garbled. Unique items, in which public libraries were still interested, were preserved as a whole and carefully scanned as books.

4.3   Structuring

Ultimately, the ad hoc-like inventory entails a pragmatic and not necessarily orthogonal structure. The archive's incremental growth leads in places to a terminology that one would not have chosen from statu nascendi. Now it is better not to change the terms that have been introduced. Inevitable latecomers may need to be matched. The archive is currently divided into 13 areas ("Bereiche", Fig. 1) according to historical arrangements.

Teilbereiche in diesem Bereich

Fachtagung Personal-Realtime-Co...	27	PEARL-Rundschau	128
PEARL-Tagung	93	Proceedings	28
Projekt Prozeßlenkung mit DV-Anla...	64	Siemens-Prozessrechner Anwende...	189
Systemsoftware	25	Anwendungsberichte	7
Berichte	2	Dissertationen	11
PEARL-Sprachbeschreibungen	5	Sprachbeschreibungen	13
Volltexte	5		

Fig. 1. The archive's topical areas

For each area the overview in Fig. 1 shows the number of documents contained. The identifier of each area leads to an explanatory header at the next level and, usually, to further subdivisions. At the lowest level one finds the digitised objects for download, together with an abstract and a citation each generated from the metadata (Fig. 2). Proceedings are represented as a whole and by individual contributions.

SPL IV: Extended Fortran IV For Process Control

Autor(en): Oerter, C.W. ☑ [DBLP] ; Peterman, G.L. ☑ [DBLP]

Zusammenfassung
Extensions to Fortran are presented to illustrate the value of building into the language facilities for known requirements.

Vollständige Referenz   BibTeX

Oerter, C. & Peterman, G., (1970). SPL IV: Extended Fortran IV For Process Control. Fifth Annual Workshop on The Use of
Digital Computers in Process Control. Lousiana State Univ.

Fig. 2. Abstract and generated citation for SPL IV

5   Content of the Archive

The heterogeneous and somewhat arbitrary stock is artificially arranged here to make it easier to read sequentially, even if the subareas overlap in terms of content. The topics are Basics, Scientific Papers, Events, Project PDV, PEARL Rundschau, and Results. The areas are presented individually. Some particular objects are presented with front matter screen shots in Figs. 3, 4, 5, 6, 7, 8, 9, 10, 11 and 12.

5.1   Basics

Language Descriptions ("Sprachbeschreibungen"). This section contains drafts for real-time and system languages. Historic highlights are BCPL, the predecessor of B and of C, by Cambridge University (1969) [1] (Fig. 3 left), RTL by Imperial Chemical Industries (1970) [4] (Fig. 3 right) and, out of competition, Fortran for Siemens 4004 (1965).

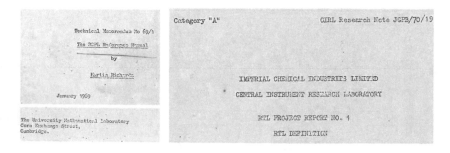

Fig. 3. Language descriptions, left: BCPL 1969 [1], right: RTL 1970 [4]

PEARL Language Descriptions ("PEARL Sprachbeschreibungen"). Here one can find company-independent PEARL language descriptions from the early days and from the time of redesign.

5.2   Scientific Material

PhD Theses ("Dissertationen"). Start and establishment of the topic real-time systems, from language definition to the development of design and test systems, was not possible without qualified scientific support. At a number of universities, this was reflected by relevant PhD theses, often in the then new discipline of computer science. The documents that can be called up here are not put together representatively, but rather come from random collections. An early example is the thesis on dynamic priority assignment (1975) [7] (Fig. 4 left), a later work (1993) [14] (Fig. 4 right) deals with mapping the ISO transaction service to real-time communication.

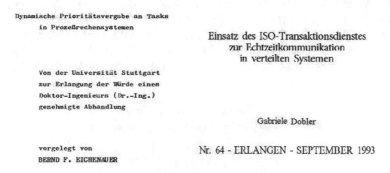

Fig. 4. Theses, left: Eichenauer 1975 [7], right: Dobler 1993 [14]

PEARL Papers ("Volltexte"). Here one finds scientifically published articles on trends in PEARL development in the initial phase (1970s) [2] and in the redesign phase (1980s).

5.3   Events

Siemens User Group ("Siemens Anwenderkreis – SAK"). The user groups of German manufacturers of process control computers, e.g. the Siemens user group (SAK), witnessed early real-time applications. The annual user conferences were organised by universities, research institutions or the manufacturers themselves. They were the fora for exchanging information about applications, software and developments. The proceedings of these conferences reflect the pioneering character of the early 1970s and report, for example, on the computer-aided control of the first German research satellite AZUR (1969/1970) [3] (Fig. 5).

J. Radünz
Deutsche Forschungs- und
Versuchsanstalt für Luft-
und Raumfahrt e.V.,

Oberpfaffenhofen

"Betriebssystem für den deutschen

Forschungssatelliten A Z U R"

Fig. 5. Control of the AZUR satellite 1970 [3]

PEARL Workshop ("PEARL Tagung"). Since the early 1980s, the
PEARL Association (PEARL-Verein, PV) held annual workshops on PEARL
in particular and real-time systems in general. This type of series was adopted
by GI in 1991. The workshop proceedings were published independently until
1988, and from 1989 on by Springer-Verlag. In 1987, Hewlett Packard published
a pioneering article on adding real-time capabilities to Unix [4] (Fig. 6).

**Adding Real Time Capabilities
to the UNIX* Operating System**

Suzanne M. Daughty

Sol F. Kavy

Steven R. Kusmer

Douglas U. Larson

David C. Lennert

Frank-Peter Schmidt-Lademann

Hewlett-Packard Company

Fig. 6. Real-time Unix by HP 1987 [4]

Personal Real-Time Computing. With the emergence of personal computers
in the mid-1980s, traditional process control computers were pushed into the
background. The PEARL Association seized this trend with annual workshops
on "Personal Real-time Computing". A contribution from 1985 about the use of
personal computers to control electrical energy supply seems ahead of its time
and somewhat strange.

Proceedings. This section of the archive contains the proceedings of an early
international conference on real-time systems held in Germany in 1972 [5]
(Fig. 7).

Computing with
REAL-TIME SYSTEMS

Volume 2

Proceedings of the
Second European Seminar
University of Erlangen – Nürnberg

Edited by
I.C. PYLE
AERE Harwell
and P. ELZER
University of Erlangen

Fig. 7. Proceedings edited by Pyle, Elzer 1972 [5]

5.4 The Project "Process Control by Data Processing" ("Prozesslenkung Durch Datenverarbeitung – PDV")

The project PDV as a part of the Federal Government's data processing programme [6] brought about the greatest development push for real-
time systems in Germany. The project's results were published in a series
of reports/development notes (Berichte/Entwicklungsnotizen) and communica-
tions/project reports (Mitteilungen/Projektberichte) by the organisation managing the project, i.e. Kernforschungszentrum Karlsruhe (KfK). These documents reflect the full breadth of the development process at that time. For example, report 76 describes the virtual assembler CIMIC used in implementations,
report 78 a portable real-time operating system for the (8-bit) microprocessor
Z80, and report 80 applications such as a cable coating process, leak monitoring in pipelines, and monitoring the wheel load conditions of rail vehicles in
rolling operation. Report 132 from 1979 presents future prospects. Here the use
of real-time systems in telecommunications is propagated [8] (Fig. 8).

PDV-E 132
September 1979

PDV-Entwicklungsnotizen

Programmiersprachen für Mikrorechner
zur Steuerung nachrichtentechnischer
Anlagen

S. Nestel, K. Schaper
SEL-Forschungszentrum, Stuttgart

Fig. 8. Real-time systems for telecommunication by SEL 1979 [8]

5.5   PEARL Review ("PEARL-Rundschau")

After the funding of the PEARL development had ceased, the development activities were bundled within the PEARL Association based at the Society of German Engineers (VDI) in Düsseldorf. Among other papers, the Association published the PEARL-Rundschau (1980 to 1982). Over time, the view of real-time systems in general broadened, as can be seen in the article [5] (Fig. 9) with a comparison of the languages Coral, Pascal, PEARL, and Ada in Volume 2, Issue 2 (1981). A good overview of the already wide range of applications can be found in the article entitled Industrial Applications of PEARL (Volume 3, Issue 1, 1982), featuring automation of soaking furnaces, electrical power distribution, data communication, and online coordinate transformation for industrial robots [6] (Fig. 10).

Comparison of Languages (Coral, PASCAL, PEARL, Ada)

H. Sandmayr

Abstract. The facilities of some languages used for realtime applications are summarized and compared. It is not intended to give a recommendation for the use of one of these languages. Instead a set of different approaches is presented which provides an overview.

The development of real-time languages is embedded in the above mentioned development. CORAL an PEARL are languages developed in the second phase mentioned above. CORAL (1964, 1966) is an attempt to combine features of ALGOL 60, FORTRAN and macroassembly languages into an efficient language suited for real-

Fig. 9. Real-time language comparison by Sandmayr 1981 [5]

Industrial Applications of PEARL

Dr. H. Steusloff, Karlsruhe (IITB)

Summary

The value of a programming language may only be determined by application experience. The language PEARL (Process and Experiment Automation Realtime Language) [1] has been desig ed to be an application programming language

Fig. 10. Industrial Applications by Steusloff 1982 [6]

5.6   Results

System Software ("Systemsoftware"). This section of the archive contains descriptions of and references to PEARL and other real-time systems, i.e. compilers and operating systems. It is striking to see how consistently the U.S. pioneer Digital Equipment had set up its commitment to PEARL [9] in the 80ies.

Reports ("Berichte"). This section contains detailed final reports of larger real-time projects funded before and after the PDV period.

Fig. 11. RSX11M operating system by Digital Equipment 1980 [9]

Application Reports ("Anwendungsberichte"). Reports on early real-time applications cannot only be found in relevant proceedings or as journal articles, but also distributed over various non-IT-oriented places. It would be difficult to find them there without detailed knowledge of authors or topic. The collection in the archive should help. It ranges from laboratory reports to regular publications. Reports on deployments in television, in high-energy physics [12] (Fig. 12), or in recording aircraft noise are striking.

> **AN INTELLIGENT MULTICHANNEL ANALYZER FOR STABILITY SUPERVISION OF PULSE HEIGHT SPECTRA**
>
> R. BARAN, R. BESOLD, A. HOFMANN, R. OLSZEWSKI and H.W. ORTNER
>
> *Physikalisches Institut der Universität Erlangen-Nürnberg, D8520 Erlangen, FRG*
>
> Received 4 November 1986 and in revised form 12 March 1987
>
> A monitoring system for pulse height spectra has been developed. It supervises peak positions and transmits data to an on-line computer. The system can be useful both for off-line analysis and during data aquisition. The hardware is based on a Z80 microprocessor. The software is written in the programming language PEARL. Two applications are presented.

Fig. 12. A multi-channel analyser in high-energy physics 1987 [12]

6   Usage

The real-time systems archive can be reached by invoking https://dl.gi.de and, then, descending to departments ("Fachbereiche") → technical informatics ("Technische Informatik") → real-time systems ("Echtzeitsysteme").

Searching by means of listing ("Auflistung") leads either to title, author, publication date or keyword. A search mask can be used to search locally or globally. A local search, e.g. for "Application", returns all articles with the searched term in the title or the text. Page layout and sorting are adjustable.

7   Experience

The real-time archive is found by internet search engines with high priority. If you search for a rather unusual term like "Project PDV", the engines will generally return the searched result collected from the archive on the first page, although the term PDV is now used elsewhere.

Unfortunately, due to its origins, the archive is steeped in the past. Newer contributions – or even historical ones – are very welcome, especially relevant PhD theses. If you have any, please contact `archiv@real-time.de`.

References

1. Baran, R., Besold, R., Hofmann, A., Olszewski, R., Ortner, H.W.: An intelligent multichannel analyzer for stability supervision of pulse height spectra. Nucl. Instrum. Methods Phys. Res. **258**(1), 91–94 (1987)
2. Brandes, J., et al.: The concept of a process-and experiment-oriented programming language. elektronische datenverarbeitung **10**, 429–442 (1970)
3. Daugthy, S.M., Kavy, S.F., Kusmer, S.R., Larson, D.V., Lennert, D.C., Schmidt-Lademann, F.-P.: Adding real time capabilities to the UNIX* operating system. In: Proceedings of the PEARL 1987, pp. 151–165. PEARL-Verein/GMA, Düsseldorf (1987)
4. Dobler, G.: Einsatz des ISO-Transaktionsdienstes zur Echtzeitkommunikation in verteilten Systemen. Dissertation, University of Erlangen-Nürnberg, Erlangen (1993)
5. Eichenauer, B.: Dynamische Prioritätsvergabe an Tasks in Prozeßrechensystemen. Dissertation, University of Stuttgart, Stuttgart (1975)
6. Nestel, S., Schaper, K.: Programmiersprachen für Mikrorechner zur Steuerung nachrichtentechnischer Anlagen. PDV-Entwicklungsnotizen PDV-E 132. KfK, Karlsruhe (1979)
7. NN: Forschungsbericht (IV) der Bundesregierung. Bundesminister für Bildung und Wissenschaft, Bonn (1972)
8. NN: RSX-11 M Real Time Operating System. Digital Equipment Corporation, Maynard (1980)
9. NN: RTL Definition. Project report, Imperial Chemical Industries Ltd., Reading (1970)
10. Pyle, I.C., Elzer, P.: Computing with Real-Time Systems, vol. 2. Transcripta Books, London (1972)
11. Radünz, J.: Betriebssystem für den deutschen Forschungssatelliten AZUR. In: Proceedings of Siemens-Prozeßrechner-Benutzertagung 1970, pp. 76–102. KFA Jülich, Jülich (1970)
12. Richards, M.: The BCPL reference manual. Technical Memorandum. The University Mathematical Laboratory, Cambridge (1969)
13. Sandmayr, H.: Comparison of Languages (Coral, PASCAL, PEARL, Ada). PEARL-Rundschau Band 2 Nr. 2, pp. 29–36. PEARL-Verein/GMA, Düsseldorf (1981)
14. Steusloff, H.: Industrial Applications of PEARL. PEARL-Rundschau Band 3 Nr. 1, pp. 35–41. PEARL-Verein/GMA, Düsseldorf (1982)

Deadlock Detection in OpenPEARL

Jan Knoblauch[(✉)]

Hochschule Furtwangen University, 79761 Waldshut-Tiengen, Germany
`jan.knoblauch@gmx.de`

Abstract. OpenPEARL is an open source build system for PEARL, a DIN-standard programming language designed for building multitasking and real-time applications. In the target domain of PEARL applications, error-free synchronization of multiple processes is of particular importance. Among other things, deadlocks pose a great risk, as they usually occur irregularly and cause the system to crash. The following describes a concept that can detect deadlocks that may occur and those that have already occurred through various approaches. It is integrated with OpenPEARL and is designed to help application developers identify, understand, and troubleshoot process synchronization errors.

Keywords: PEARL · OpenPEARL · deadlock · static analysis · dynamic analysis · control flow graph · resource allocation graph

1 Deadlock

Deadlocks represent a danger in application development that should not be underestimated. They occur when each process of a set of processes waits for an action of another of these processes. Often these actions are the release of resources such as files, synchronization locks, interfaces or peripherals. Deadlocks are usually dependent on several factors that are not under the control of the application. Due to external factors such as non-deterministic scheduling between processes and different latencies in network and hardware accesses, deadlocks can occur very irregularly. Deadlocks are usually not detectable by common testing methods such as unit and module tests. The occurrence of deadlocks may differ in development, test and production environments, for example, if factors such as system performance and load are different. If a deadlock occurs regularly, it is probably discovered and fixed during development or testing. But if the occurrence of a deadlock depends on several rather unlikely input data, function calls, external applications or services, the problems may not be detected.

In a deadlock situation, all involved processes are not executed any further, and a deadlock exists that normally cannot be resolved in an automated way. If a deadlock thus occurs, the system can usually not continue to work, but must be restarted. External intervention is needed, whether manual or from a monitoring system. Deadlocks are therefore considered critical faults that are difficult to find through testing, and often lead to failures when they do occur. Especially with high reliability requirements, it must therefore be possible to systematically exclude deadlocks.

For a deadlock to occur, [2] states that four conditions are sufficient.

H. Unger and M. Schaible (Eds.): Real-Time 2022, LNNS 674, pp. 13–22, 2023.
https://doi.org/10.1007/978-3-031-32700-1_2

1. No Preemption: Resources are released exclusively by the processes themselves
2. Hold and Wait: The processes already have resources and still need more
3. Mutual Exclusion: Only one process can access a resource at the same time
4. Circular Wait: A circular dependency exists between at least two resources of two processes

For a deadlock to occur, it is sufficient if these four sufficient conditions are met. The conditions imply a deadlock, but there is no equivalence between these conditions and the occurrence of a deadlock. Therefore, deadlocks can also occur if not all of these conditions are met.

Several ways exist to avoid deadlocks during application development. However, it depends on the developers whether and how carefully these mechanisms are implemented. In [1], deadlock treatment strategies are classified into the three following categories.

1.1 Deadlock Prevention

The deadlock prevention approach describes the prevention of necessary conditions for deadlocks on a logical level. According to [4], it is sufficient to enforce one of the following approaches.

The condition "Hold and Wait" is not satisfied if all resources must be requested simultaneously, which is described by the approach "Simultaneous Locking". A process must request all resources it may need at the same time and cannot request resources again until it has released all occupied resources. However, this severely limits parallelism when a process requires one resource for only a short period of time and another resource for a comparatively long period of time. According to the principle of simultaneous locking, the process would have to occupy both resources at the same time, even if it does not use the resource required for a short time until much later. During this time, this resource would also not be available to the other processes and would therefore remain unused. This prevents the optimal allocation of resources and therefore severely limits parallelism.

In contrast, the approach "Ordered Locking" ensures that all resources can only be requested in a specific order. For this, a sequence must be formed over all resources of a system. A process can only request resources that have a higher order in this sequence than the resources that the process already occupies. In order to occupy a lower-order resource, the process must first release its occupied higher-order resources. This means that no cyclical dependencies can arise between processes and their operating resources. Thus, the condition "Circular Wait" is no longer fulfilled.

1.2 Deadlock Avoidance

Deadlocks can also be actively avoided during the execution of an application, which is called "Deadlock Avoidance". Thereby, the scheduling of the processes is controlled in such a way that there is always a sequence in which all upcoming resource requests can be executed.

A method to avoid deadlocks was introduced by Edsger W. Dijkstra in 1965 as "Bankers Algorithm" and is described in [4]. Adapted from a banker who grants loans of his customers, it controls the allocation of resources. At the beginning each process must be known, and which resources it will need during the execution. The algorithm describes that before each resource allocation, it is checked whether the application is still in a safe state afterwards. If this is the case, the resource is allocated, otherwise the allocation is delayed out, even if it would be feasible at the current time. A state is considered safe if a scheduling order is possible in which all requested resources can be allocated without jams. If this is not the case, the state is unsafe. Depending on the scheduling, allocations and releases of the processes, deadlocks can occur in an unsafe state.

By executing or delaying possible resource allocations through the deadlock avoidance depending on the state, the use of it actively changes the behavior of resource allocations and thus the runtime behavior of the entire application. The PEARL definition is to allocate requested resources when they are available. However, by delaying possible resource allocations through deadlock avoidance, this would affect the runtime behavior of the application. This approach will therefore not be examined further.

1.3   Deadlock Detection and Resolution

The third category of deadlock treatment strategies is the detection of deadlock situations that have occurred during execution, as well as their resolution. To resolve a deadlock, a resource must be withdrawn from a participating task. Basically, the resource can only be revoked if the operations that the task has applied to the resource are undone. In order to be able to revoke a resource from an PEARL task according to this principle, the task's operations on the resource must be able to be undone. In this context, the synchronization mechanisms of PEARL do not represent the actual resources, but in the field of embedded systems, for example, these resources are motors and actuators. The synchronization mechanisms merely serve to control which task may use which resource. Even if the generated actions of a task on an actuator were logged and undone in the event of a problem, this would not be possible in a generally valid manner, and certainly would not make sense. For this reason, it is generally not possible to withdraw a resource from a task, and therefore deadlocks that have occurred cannot be resolved automatically in PEARL (and other general-purpose languages) even at the logical level.

2   OpenPEARL

The standardized [1] programming language PEARL ("Process and Experiment Automation Realtime Language") provides convenient support for requirements

---

[1] DIN 66253:2018-03.

from the areas of multitasking and real-time capability. OpenPEARL is an open source project and provides a compiler and runtime environment for Open-PEARL programs for various target systems (Fig. 1).

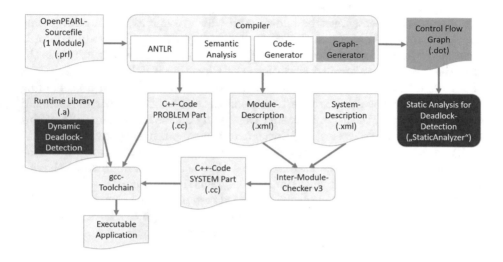

Fig. 1. Extended architecture of OpenPEARL

To provide the most helpful support for detecting deadlocks and developing deadlock-free applications, the concept includes static deadlock analysis during compilation and dynamic deadlock detection during execution. For this purpose, the architecture of OpenPEARL has been extended to include the components highlighted in color. The compiler generates a control flow graph in DOT format during compilation. The static analysis is implemented in a standalone application that analyses this control flow graph. The dynamic analysis is integrated into the runtime environment.

2.1   Deadlock-Relevant Operations

The implemented analyses address OpenPEARL statements that can potentially lead to, or prevent, a deadlock situation. These instructions are limited in OpenPEARL to the "SEMA" and "BOLT" type access functions, and are called deadlock operations. All operations on SEMA and BOLT variables are performed atomically in OpenPEARL for all variables specified in the operation.

SEMA. The SEMA type provides counting semaphores, binary semaphores or mutexes are not available. Three operations can be performed on SEMA resources: "REQUEST", "RELEASE" and "TRY". By means of "REQUEST" the specified SEMA resources are occupied, and they are released by means of "RELEASE". The value of the semaphore always changes by only one in each single step, but in an operation the same resource can be specified multiple times to be able to increase or decrease the value atomically multiple times on the logical level. The operation "TRY" can be used to try to allocate a SEMA resource. Since a "REQUEST" operation is implicitly performed when occupancy is possible, the "TRY" operation need not be further explored in the dynamic analysis. In the static analysis, the "TRY" operation is not considered.

BOLT. Using BOLT resources, the users of a resource can be divided into two groups, the exclusive users and the concurrent users. In general, exclusive users perform write accesses to protected data and concurrent users perform read-only accesses. For write accesses, it is important to grant access to only a single writing user at a time. Reading users should also be locked during write operations to prevent incompletely modified data from being read. However, if the data is not modified, any number of reading users can access the data simultaneously. Using "ENTER" and "LEAVE", reading users occupy the BOLT resource and release it. Exclusive allocation and release is performed by means of "RESERVE" and "FREE". A BOLT resource can assume three states. If it is occupied by an exclusive user, it is "locked". If it is occupied by one or more reading users, its state is "not lockable". If it is occupied neither by an exclusive user nor by reading users, it is "lockable". For BOLT variables, exclusive assignments always take precedence over shared access. Non-exclusive occupancy of BOLT resources is countable. Since typical deadlocks due to the use of BOLT resources can only occur due to exclusive occupancy, the dynamic analysis is limited to their "RESERVE" and "FREE" operations.

3 Static Analysis

The goal of static analysis is to notify the application developer of problems that could cause deadlocks before the application is executed. In addition, static analysis should be able to rule out deadlocks during execution if an application does not meet necessary conditions for deadlocks. Because static analysis cannot fully recreate and examine the runtime behavior of a (multithreaded) application, it generally cannot predict deadlocks with certainty, but can only indicate that deadlocks may occur at certain points.

The static analysis is based on the control flow graph that is generated during compilation. This control flow graph contains, in addition to all statements, various meta-information such as information about tasks, procedures, and SEMA and BOLT resources. In addition, the graph contains the code position of each instruction consisting of file name and line number. In order to analyse this graph, a number of modifications and simplifications are made to it. For example, all operations that are not deadlock relevant or unreachable are removed. For the static analysis, a graph remains on the basis of which all possible processing paths can be determined for all tasks. For example, it must be determined which task can execute which procedures (recursively) and which other tasks it can activate. Multiple paths are created by conditions, case branches, and loops. The static analysis does not evaluate values and ranges of values, which is why it does not determine, for example, how many times a loop body could be executed. Instead, it only considers the two cases where the loop is not run or is run exactly once. Based on the execution paths, static analysis performs some checks, such as looking for mismatched allocations at the block, task, and module levels. If, for example, a SEMA resource is only occupied by a task and not released, there is possibly a mistake. But if a task exists that only releases the same resource, this behavior may be correct. A typical scenario with possible deadlock risk exists, for example, if several processing paths of a task release a resource and a single one does not release it, for example, because a procedure was exited early.

In addition to a large number of these checks, the static analysis performs two central inspections to exclude the fulfillment of the sufficient conditions "Hold and Wait" and "Circular Wait". Whether "Simultaneous Locking" is met can be determined by checking for all processing paths of all tasks whether resources are requested while the task is already occupying resources. On the other hand, to check whether "Ordered Locking", which is much more practical, is adhered to, an occupancy sequence must be formed for all deadlock-relevant resources. This could be specified explicitly, for example via annotations in the OpenPEARL code, but this would result in increased maintenance effort and possible errors. Instead, a topological sort is formed using the resource assignments of the task execution paths until all resources are included. Topological sorting based on depth-first search as described in [3] is used to obtain topological sorting even for cyclic graphs. Subsequently, all processing paths are checked to see if the order formed is followed. Violations indicate deadlocks that may occur. If, on the other hand, there are no violations, no deadlocks can occur, in which a set of tasks waits for resources that are each occupied by other tasks of this set.

Static analysis is an optional step of the build process that is not intended to abort the process if problems are detected. The area of use of SEMA and BOLT resources is diverse and not every type of use can be detected and taken into account in the static analysis. If the analysis prevented compilation, it would significantly limit application developers in the case of incorrectly detected but correctly implemented applications. Since it cannot accurately recreate runtime behavior, and this depends on many external factors, deadlocks in general cannot be predicted with certainty statically.

Since the control flow graph contains various meta-information, static analysis can produce very detailed outputs. In addition, all outputs are classified into three categories: information, warning, and error. Problems that definitely lead to a runtime error during execution, such as the multiple exclusive allocation of a BOLT resource, are classified as errors.

4   Dynamic Analysis

The goal of dynamic analysis during the execution of the application is to be able
to detect and describe deadlocks that have occurred. Before an application is executed, static analysis can be used to investigate whether deadlock situations can
occur during execution. However, the fact that an application is currently stuck
in a deadlock situation can only be detected at runtime. If an occurred deadlock
can be detected immediately during execution, it can be reacted to instead of
continuing to execute the blocked application. An obvious reaction to a deadlock that has occurred is to terminate and restart the application. Tasks not
involved in a deadlock could be terminated in a controlled manner beforehand
and the application thus terminated correctly, at least in part. If the application
is then restarted, the downtime can be kept to a minimum. Without deadlock
detection, manual intervention is required because a deadlock blocks an application endlessly. However, dynamic analysis should not only detect deadlocks, but
also be able to describe them as meaningfully as possible. Deadlocks that have
occurred should thus be understandable for application developers so that they
can correct the underlying logic errors.

4.1   Resource-Allocation-Graph

A deadlock occurs when there is a cyclic dependency between different resources
and consumers (tasks, processes, threads, but also programs or computer systems). According to [4], such cycles can be recognized by an "Resource-
Allocation-Graph", also called "Wait-for Graph".

The resource allocation graph is a directed graph. Each resource and each
task is represented as a node. The graph is bipartite because the nodes can be
divided into two disjoint subsets, the resources and the tasks. There are no edges
within these subsets. In [4], the convention is to represent the nodes of resources
as square and those of tasks as round, as shown in Fig. 2.

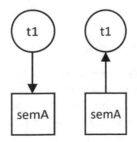

Fig. 2. Resource allocation graph. Left: Task "t1" requests resource "semA", right:
"t1" occupies "semA"

If a task now wants to occupy a resource and this is available, a directed
edge is created from the resource to the task. This resource is thus assigned

to this task, it occupies it (occupation edge, left). However, if this resource is already occupied, the request is created as an edge from the task to the resource (request edge, right). Failure to grant a request for a resource causes the task to wait until the resource is allocated to it. The task cannot perform any other operations during this time.

The resource allocation graph can be updated during execution whenever resources are requested and released. As soon as a cyclic dependency between resources and tasks occurs, a cycle also exists in this graph. Since removing an edge cannot create a new cycle, it is only necessary to check whether a cycle exists each time an edge is created. In Fig. 3, a cyclic dependency exists between tasks "t1" and "t2" and resources "semA" and "semB". Thus, a deadlock has occurred.

In order to be able to describe and reconstruct deadlocks that have occurred as accurately as possible, various meta-information is assigned to each edge in the resource allocation graph. These contain, among other things, the position of the instruction, the task that performed the operation, and the timestamp of the execution, in order to be able to generate a chronological sequence of operations in the event of a deadlock occurring.

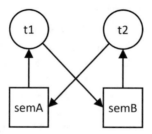

Fig. 3. Circular dependency between tasks "t1" and "t2" and resources "semA" and "semB": Deadlock occurred

4.2   Definite and Temporary Deadlocks

Deadlock detection using the resource allocation graph assumes that only the task occupying a resource can release it ("No Preemption"). If the tasks of a cycle in the graph are blocked by waiting for other resources, they will never release the resources and thereby also wait endlessly for other resources. In this case, an unresolvable deadlock has occurred. However, the deadlock-relevant resources in OpenPEARL can be released by tasks that do not occupy them themselves. For this reason, a deadlock can be released by uninvolved tasks.

A cycle in the resource allocation graph thus does not always represent an endless deadlock in the present case, but may possibly resolve itself further down the line. For this reason, when a cycle is detected during execution, it must be checked whether an uninvolved task releases one of the involved resources.

By including information about deadlock operations in the compilation, the runtime environment knows all possibly executable operations of an application. If a cycle is now discovered, it is only necessary to search for an operation which releases an involved resource and can still be executed by an uninvolved task. If a suitable operation is available, it cannot be excluded that the jam is resolved in the further process. Only if no suitable operation can be found, an endless deadlock is present with certainty. Therefore, dynamic analysis distinguishes between definite and possibly resolvable deadlocks (temporary deadlocks). If temporary deadlocks occur regularly, this indicates uneven utilization of subsystems and should be investigated further.

A possible extension that would allow the runtime environment to better decide whether a deadlock can be resolved at some point or not is the control flow graph. If the control flow graph of all tasks and procedures is available to the runtime environment, it could record the executed statements of the tasks in this graph. From this, it would be possible to derive whether certain instructions are still attainable in the further course of the application at all. In addition, it could be used to detect whether currently inactive tasks might be reactivated by other tasks or events, or whether this can be ruled out. This further information would make it possible to detect more deadlocks as final, if it can be checked whether uninvolved tasks can still release resources involved in a deadlock in the further course. This extension would allow significantly more endless deadlocks to actually be detected as endless, although not all, since deadlock resolution may still depend on external factors such as interrupts. The disadvantage of this approach is increased resource consumption, which would be disproportionate to the benefit, depending on the use case. Each executed instruction of each task would have to be captured in the control flow graph, and the investigation of whether resolution is possible would also be more extensive. Therefore, this extension is not investigated further.

5   Conclusion

In order to be able to exclude deadlocks, especially in critical software, static analyses and concepts are necessary to be able to exclude deadlocks and, in addition, also undesirable behavior of the application. The static and dynamic deadlock detection form a good assistance and meaningful extension of the Open-PEARL build system. The developed analyses are also suitable for other languages that provide multitasking and comparable synchronization mechanisms.

5.1   Evaluation of Static Deadlock Detection

The static analysis described reliably detects inconsistencies and anomalies and generates comprehensible and helpful messages. Cyclic dependencies can be ruled out by fully examining whether a uniform occupancy sequence can be formed across all resources used in a module, which is adhered to at all points. By checking whether deadlock-relevant resources are used sensibly at module, task and path level, some programming errors can be detected. Static analysis can help application developers develop deadlock-free code. Even if logic errors in the application as well as the interaction of several tasks still cannot rule out the possibility that an application does not behave as desired, numerous critical errors are detected and reported.

5.2   Evaluation of Dynamic Deadlock Detection

Deadlocks can be reliably detected during execution by the described extensions in the runtime environment. Without this detection, depending on the use case, it is not possible to determine whether an application is still working or blocked. Already the recognition that a problem has occurred in an application, as well as the detailed description of the problem, represent an enormous added value. This enables a quick response to the deadlock and helps reconstruct the processes that led to the problem. Detecting potential deadlocks during execution can help identify common and regular blockages in an application that may be mitigated.

References

1. Bruns, M.: Entwurf von Steuerungen für komplexe Transportsysteme, Rheinisch-Westfälische Technische Hochschule Aachen (1983)
2. Coffman, E., Elphick, M., Shoshani, A.: System deadlocks. ACM Comput. Surv. **3**, 67–78 (1971)
3. Cormen, T., Leiserson, C., Rivest, R., Stein, C.: Algorithmen - Eine Einführung, Oldenbourg (2010)
4. Tanenbaum, A.: Moderne Betriebssysteme. Pearson Deutschland GmbH (2009)

Development of an Authentication Method for Time-Sensitive Networking Using FPGAs and Delay-Based Physical Unclonable Functions: A Research Plan

Sinan Yavuz[1,2(✉)], Edwin Naroska[1], and Kai Daniel[2]

[1] Niederrhein University of Applied Sciences, 47805 Krefeld, Germany
Sinan.Yavuz@hs-niederrhein.de
[2] University of Siegen, 57076 Siegen, Germany

Abstract. Real-time data transmission and associated security of technical devices in industrial environments is an important key technology due to the rapid development of distributed systems. Especially, security and protection of process-related information on lightweight networked devices is very important, but also a major challenge because of the limited hardware resources of such devices. Encryption techniques by using traditional cryptographic approaches require a large number of computations, which are often not suitable for small devices and makes the implementation infeasible or inefficient. To address this issue, Physical Unclonable Functions (PUFs) may be used as a lightweight, fast, and low-cost security component on ASICS or FPGAs. In this paper, investigations for a further research project in the field of secure real-time transmission of data with Time-Sensitive Networking and PUFs are presented. For this, the components, possibilities, and security analysis of PUFs and TSN are introduced and an overview of the hardware architecture used within the research project is presented.

Keywords: PUF-based protocols · Time-Sensitive Networking · Hardware security · IoT

1 Introduction

Fast communication networks have a significant influence on the development of time-critical applications. Achieving high bandwidths and exchanging data with low latency is mandatory for time-critical applications. In Industry 4.0, networked IIoT (Industrial Internet of Things) devices such as sensors, actuators, or controllers are used to control or monitor production steps and are developed to perform specific tasks such as measuring machine conditions like ambient temperature, vibration etc. A real-time communication network can hence enable rapid detection of dangerous situations in vehicles or machines, or real-time monitoring and control of industrial plants in order to optimize production steps and intervene quickly to important events.

© The Author(s), under exclusive license to Springer Nature Switzerland AG 2023
H. Unger and M. Schaible (Eds.): Real-Time 2022, LNNS 674, pp. 23–32, 2023.
https://doi.org/10.1007/978-3-031-32700-1_3

To provide real-time data transmission in time-critical domains, the extended ethernet standards IEEE 802.1 Time-Sensitive Networking (TSN) can be used, which assures deterministic communication. The standards can be used in different areas such as Audio Video Bridging, Industrial Automation, vehicles etc.

Security and the prevention against attacks are indispensable factors for mission-critical applications. Especially, when transmitting sensitive data such as images, sensor values, parameters, etc. There are different attacks techniques that can be performed against devices or on networks such as physical, modeling, or spoofing attacks. As an example, an attacker can spoof data frames by manipulating the ARP cache table or clone hardware to impersonate as a legitimate host on a network.

Implementation of security applications especially on small devices such as in IIoT are a major challenge due to the limited resources of the devices. Lightweight devices do not have a large number of hardware resources, processing elements or large memory and are often powered by batteries. Hence, it is also very difficult to implement computational and memory intensive security techniques. As a result, these devices often have a lightweight, minimal or no security functions [1]. In this case, the implementation and the usage of silicon-based Physical Unclonable Functions (PUFs), a hardware-based security primitive, which are implemented on hardware, can be used to increase the security of a lightweight device. Silicon-based PUFs exploit the characteristics of integrated circuits (ICs), which are produced by fluctuations in the manufacturing process. Due to the different fluctuation on the manufacturing process, a PUF will generate a different unique output signal (Response) for a specific input (Challenge) if its implemented on different devices. Therefore, these input-output pairs (called Challenge-Response-Pairs) can also be used to authenticate or identify device to prevent cloning attacks. In comparison to conventional cryptographic security technologies, PUFs are lightweight, efficient, and fast. In addition, there are no standardized security techniques of PUFs, which makes it as an attractive security component on lightweight devices.

This paper gives you an overview of our research plan in the field of secure transmission of data between authenticated devices with low latency and suitable for lightweight devices by using TSN and PUFs. For this purpose, a brief introduction to TSN and PUFs are presented. In addition, the security analyses and prevention techniques of two recent PUF based authentication protocols are also described in this paper.

This paper is structured as follows: In the first section, an introduction to Time-Sensitive Networking is presented. The next section introduces Physical Unclonable Functions and analyses the security features of two recent PUF-based authentication protocols. Next, Sect. 4 presents the hardware architecture and steps for future investigations. A conclusion is drawn in Sect. 5.

2 Time-Sensitive Networking

In time-critical networks, fast processing and exchange of data between connected devices is a key issue. Due to the extension of real-time capabilities of IEEE 802.3 Ethernet, TSN is an attractive alternative for use in time-critical networks. TSN is the name of a set of standards and projects developed by the TSN Task Group of the 802.1 Working Group, which are used to ensure deterministic services through IEEE 802 networks [2] and the transmission of data packets with low latency. Mechanisms for the reservation of resources and bandwidth for time-critical frames are also considered.

The standards are used in the data link layer, the seconds layer of the Open Systems Interconnection (OSI) architecture model. In this layer, critical functions for the transmission of data over network are performed. The functions encapsulate/decapsulate packets received/transmitted by the physical layer, perform addressing mechanisms for the transmission from source to destination device, synchronization of data frames, or control mechanisms for the received/transmitted packets.

The TSN standards can be categorized in four categories Time-Synchronization, Reliability, Latency and Resource Management, which are briefly introduced in the following.

2.1 TSN Components

Time Synchronization. In real-time communication networks as well as TSN, synchronization of the time of each participant in a network is an essential factor due to the transmission and processing of data with low latency.

Time Synchronization protocols such as IEEE 802.1AS-2011 allow all TSN devices in a network to synchronize to a global master clock, called grandmaster (GM). For synch, the Best Master Clock Algorithm (BMCA) can be used to select a network root timing reference [3]. Additionally, BMCA is also used to define a new grandmaster, in case of a failure of the active grandmaster [4]. Synchronization of time is performed by using Precision Time Protocol (PTP) and basically achieved by recording timestamps and transmitting of time synchronization packets between a master and a slave. Based on the recorded timestamps, the slave can calculate the offset and delay time, thus synchronize the clock.

Reliability. Enabling of high reliability is indispensable for time-critical applications. An incomplete delivery or faults during transmitting of data frames from source to destination can cause significant damages.

For increasing reliability and hence increasing/ensuring arrival of data packets to a destination device, the standard IEEE 802.1CB – Frame Replication and Elimination for Reliability can be used. In this standard, the sender duplicates the data frames, which are addressed to a destination device, and sends the packets over multiple paths to the device. The devices in the network identify and eliminate the redundant packages [5].

Latency. Transmission of packets with low latency is also essential for time-critical networks. This part regulates and distributes the data packets for communication with low latency. Each packet has a certain transmission time, which depends on a few milliseconds or lower, that should not be exceeded. This will be guaranteed the transmission of data between two nodes in a defined latency.

An optimized scheduling, forwarding and queuing procedures are important components for enabling communications with low latency. To this end, there are various mechanisms such as IEEE 802.1Qbv: Enhancements for scheduled traffic, or IEEE 802.1Qbu: Frame Preemption, which can be used. As an example, the standard IEEE 802.1Qbv assigns priorities and uses time synchronization to transmit TSN packets on a scheduled-driven communication [6]. Mechanisms for the interruption of non-time-critical data frames to transmit time-critical packets are defined by IEEE 802.1Qbu.

Resource Management. Another major TSN component to ensure deterministic communications is Resource Management. This component basically organizes mechanisms for the reservation of bandwidth for time-critical data frames. As an example, IEEE 802.1at - Stream Reservation Protocol (SRP) is used to allow configurable reservation and the management of bandwidth for time critical packets such as in industry, vehicles, or transmitting high quality video/audio streams.

2.2   Security Analysis

As already described, security of the data link layer is an important challenge. On the other hand, the protocols on the data link layer are not designed with security features and thus do not contain security mechanisms [7] which increases the probability of an attack. An effective prevention technique to increase the security is to authenticate nodes in a network and allow data transmission only between authenticated devices/nodes [8].

There are various attack techniques that can be performed against data link layer such as ARP Spoofing, MAC flooding, Port stealing or Man-in-the-middle attacks. Thus, attacks on the data link layer can also compromise the TSN standards. In addition, TSN standards also have weaknesses that must be taken into account. A successful attack on time-sensitive networks can compromise the real-time capabilities and can interrupt the network.

One of the most important parameters in time-critical networks, which have to be protected, is time [3]. Accessing and tampering of the master clock or tampering of time-synchronization packets can interrupt the communication of all the devices in the network. To increase the security, unauthorized joins from unknown devices should be detected and rejected.

Another issue is that TSN components do not necessarily support all TSN standards. If networked devices support different TSN standards, real-time transmission of data between two devices cannot be guaranteed. Likewise, an unauthorized networked TSN device can flood the network with randomized data with high priority.

Exploitation of the vulnerabilities of the TSN standards are also important attack vectors that can affect the communication. Hence, these kind of attacks also have to be taken into account. The attackers can use the weaknesses to manipulate important configurations, sequence numbers, networked specified parameters or delaying time-critical packets [3]. Therefore, detection techniques against spoofed timestamps [9] or securing critical standards by detecting anomalies [10] are essential to increase the security.

3   Physical Unclonable Functions

A silicon based PUF, which can be implemented on FPGAs or ASICs, is basically a circuit that generates unique signals such as biometrical fingerprints by evaluating variations in the manufacturing process such as delay time [11]. The generated unique signals can be used in various security applications such as on-chip key extractor [12] or to authenticate a networked device, which is indispensable to increase security and prevent influences of invalid nodes on the network.

As with any circuit, PUFs are dependent on various physical fluctuations such as temperature, electromagnetic fields, as well as aging processes. These fluctuations can affect the specific characteristics of hardware components, which can affect the output of a PUF. Thus, a major challenge in the development of PUF-based authentication protocols is to generate strong, stable, and continuable PUF outputs under physical fluctuations. As a countermeasure to physical influences, Error Correction Code (ECC) [12] can be used to increase the reliability of PUFs, which is not discussed in this paper.

3.1   PUF-Based Authentication Protocols

Authentication protocols verifies devices to ensure transmission of data between trusted hosts only. Therefore, security properties of PUFs are attractive for authentication protocols due to the generating of device-unique signals. The basic structure and the phases of an PUF-based authentication protocol is shown in Fig. 1.

After receiving a request for registration, the server generates a random challenge and sends it to the device. The device receives the challenge and outputs a response by using its PUF. After generating a response, the device sends the generated response back to the server, where the data is stored in a database. In the authentication phase, the device sends a request for the authentication similar to the first phase. The server tries to fetch the CRPs for the device ID. If there no entry in the database, the device is not already registered or has been cloned (which is an attack). If a CRP for the specific device exists, the server sends the challenge to the device. The device receives the challenge and generates an output, which is send back to the server. Finally, the server compares the received with the stored response. The connection is established if the responses match.

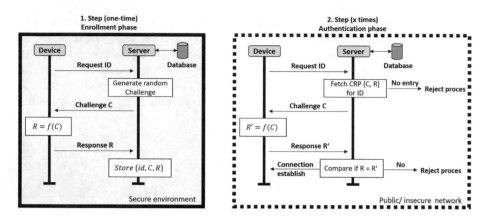

Fig. 1. Enrollment and Authentications phase of PUF-based authentication protocols

An PUF-based authentication scheme for lightweight medical devices/things is presented in [13]. In this proposed authentication protocol, a PUF is implemented on a IoMT (Internet of Medical Things) device and in a server. The registrations phase is shown in Fig. 2. The server generates a challenge C1 and outputs a response R1 by using its PUF structure. The response is sent to the IoMT device, which generates a response R from the response R1 and sends it back to the server. These steps are repeated by the server. Then, the third generated response R2 will be hashed and stored with the first challenge C1 in a database. In the authentication step, the server obtains the first challenge C1 from the secure database and mirrors the steps of the registration phase. The device is authenticated if the encrypted responses in the database are the same.

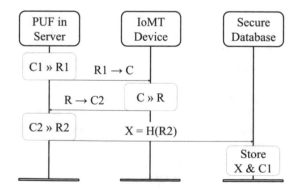

Fig. 2. Proposed enrollment step in [13]

During registration as well as authentication, the IoMT device only generates a response for a transmitted challenge. There are no computational excessive encryption functions implemented, which makes it possible to implement these authentication protocol on lightweight devices. However, the protocol has a significant weakness by transmitting the Challenge-Responses in cleartext. In an untrusted network, the CRPs can be received by an attacker and could be used to interrupt the authentication of a legitimate host. In addition, there are no techniques for re-using of CRPs in [13].

In [14], an PUF-based authentication protocol for lightweight IoT devices by using Machine Learning (ML) algorithms is proposed. In the enrollment phase the verifier stores the ID, MAC address, and a pre-trained ML model of the PUF. Then, in the authentication step the device sends the device ID as an authentication request. After receiving the request, the verifier checks the MAC address and fetches the pre-saved PUF model from the database for the specific device. Furthermore, the verifier generates a random challenge and sends it to the device. The devices use the implemented PUF structure to generate a response, which will be sent encrypted back to the verifier. The verifier mirrors these steps by using the PUF model and compares the generated secret with the transmitted secret from the device. The device is authenticated if these secrets are equal.

In this approach, it is required that the verifier models the PUF structure for each device. For this purpose, a set of CRPs for the training of the ML algorithm is needed, which means a high memory and computation overhead. The advantage of this protocol is that the response is obfuscated and sent encrypted over network. This will prevent the authentication step against Man-in-the-middle attacks.

4 Future Investigations

In this research plan, a lightweight and fast PUF-based authentication protocol is investigated to enhance the network security and ensure secure real-time transmission of data for mission-critical applications. In addition, implementation of energy efficient and low-cost hardware architecture suitable for small networked systems are also addressed. Furthermore, measurements in the field of security analysis, power consumption and delay time are also examined in this research project. For this, several FPGAs are connected in a network.

The basic structure of a single node is shown in Fig. 3.

The PS (processing system) side of the FPGA is used to execute process-related functions. Here, various security options such as secure boot or trustzone are enabled to increase the security. The FPGAs consist of an TSN IP-Core to ensure data exchange by using TSN standards. Also, a robust and power efficient PUF structure is implemented to generate hardware specific keys. This PUF structure is connected to an ECC module to generate secrets with high reliability under physical fluctuations. An additional module to detect anomalies in case of an attack against hardware and/or TSN standards is also implemented. In this case, different parameters such as temperature, voltage are measured and the output of the PUF is analyzed. If an attack is detected or an anomaly occurs, the device disables critical components and recover to a safety condition to secure data against attacks.

To ensure real-time data exchange only between authenticated devices, secret keys by using network parameters such as clock time, priority, or bandwidth in combination with PUF as a hardware specific identifier are generated. This will make sure that tampering on the PUF structure will be detected by comparing the CRPs. In addition, this ensures that the same TSN standards are used for communication.

The next step will focus on implementing of efficient and strong PUF architectures on an FPGA. Also, implementation of an TSN IP Core is also considered. Furthermore, fast generation of secret keys by using PUFs and network parameters will be investigated.

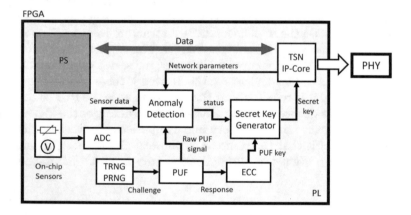

Fig. 3. Proposed architecture of a single device

5   Conclusion

Investigation of lightweight authentication techniques for small devices is an important challenge in the research. As already described, an attack to smart networked systems can lead to a dangerous situation. PUF-based authentications schemes are attractive alternatives due to the lightweight, low-energy and computational efficient structure. Also, PUFs are unclonable, unpredictable and could not reproduced which are appropriate features for security applications.

In this paper, an overview to PUF-based authentication protocols is presented. We introduced the general structure and phases of a basic authentication procedure and present two different approaches in this field. In addition, a brief introduction to TSN and PUF structures are also presented. Furthermore, the hardware architecture and next steps for further investigations were also presented in this paper.

Recent PUF-based authentication protocols are often not suitable for lightweight devices due to the usage of compute- and resource intensive encryption techniques. Computationally intensive encryption technologies have a high processing time that could lead to problems in time-critical applications. Additionally, the protocols have a single point of failure. After a successful attack on a server, the security of all the devices in a network are limited and a secure transmission of data cannot be guaranteed. Hence, to ensure secure data transmission with low latency between lightweight devices, more investigations in PUF-based authentications protocols are required.

References

1. Yoon, S., Kim, B., Kang, Y., Choi, D.: PUF-based authentication scheme for IoT devices. In: 2020 International Conference on Information and Communication Technology Convergence (ICTC), pp. 1792–1794 (2020). https://doi.org/10.1109/ICTC49870.2020.9289260
2. Farkas, J.: Introduction to IEEE 802.1 (2017). https://www.ieee802.org/1/files/public/docs2017/tsn-farkas-intro-0517-v01.pdf#page=27
3. Ergenç, D., Brülhart, C., Neumann, J., Krüger, L., Fischer, M.: On the security of IEEE 802.1 time-sensitive networking. In: 2021 IEEE International Conference on Communications Workshops (ICC Workshops), pp. 1–6 (2021). https://doi.org/10.1109/ICCWorkshops50388.2021.9473542
4. Samii, S., Zinner, H.: Level 5 by layer 2: time-sensitive networking for autonomous vehicles. IEEE Commun. Stand. Mag. **2**, 62–68 (2018)
5. Yang, X.L., Scholz, D., Helm, M.: Deterministic networking (DetNet) vs time sensitive networking (TSN) (2019)
6. Lo Bello, L., Steiner, W.: A perspective on IEEE time-sensitive networking for industrial communication and automation systems. Proc. IEEE **107**(6), 1094–1120 (2019). https://doi.org/10.1109/JPROC.2019.2905334
7. Altunbasak, H., Owen, H.: An architectural framework for data link layer security with security inter-layering. In: Proceedings 2007 IEEE SoutheastCon, pp. 607–614 (2007). https://doi.org/10.1109/SECON.2007.342975
8. Annapurna, A.: Data link layer-security issues. Int. J. Comput. Sci. Eng. Technol. (IJCSET) **4**, 1009–1012 (2013)
9. Moussa, B., Robillard, C., Zugenmaier, A., Kassouf, M., Debbabi, M., Assi, C.: Securing the precision time protocol (PTP) against fake timestamps. IEEE Commun. Lett. **23**(2), 278–281 (2019). https://doi.org/10.1109/LCOMM.2018.2883287
10. Luo, F., Wang, B., Fang, Z., Yang, Z., Jiang, Y.: Security analysis of the TSN backbone architecture and anomaly detection system design based on IEEE 802.1Qci. Secur. Commun. Netw. (2021)
11. Hiller, M., Kurzinger, L., Sigl, G.: Review of error correction for PUFs and evaluation on state-of-the-art FPGAs. J. Cryptogr. Eng. **10**, 229–247 (2020). https://doi.org/10.1007/s13389-020-00223-w
12. Usmani, M.A., Keshavarz, S., Matthews, E., Shannon, L., Tessier, R., Holcomb, D.E.: Efficient PUF-based key generation in FPGAs using per-device configuration. IEEE Trans. Very Large Scale Integr. (VLSI) Syst. **27**(2), 364–375 (2019). https://doi.org/10.1109/TVLSI.2018.2877438

13. Yanambaka, V.P., Mohanty, S.P., Kougianos, E., Puthal, D.: PMsec: physical unclonable function-based robust and lightweight authentication in the internet of medical things. IEEE Trans. Consum. Electron. **65**(3), 388–397 (2019). https://doi.org/10.1109/TCE.2019.2926192
14. Yilmaz, Y., Gunn, S.R., Halak, B.: Lightweight PUF-based authentication protocol for IoT devices. In: 2018 IEEE 3rd International Verification and Security Workshop (IVSW), pp. 38–43 (2018). https://doi.org/10.1109/IVSW.2018.8494884

Taking Real-Time and Virtualization to Open Source Hardware

Alexander Schönborn$^{(\boxtimes)}$, Robert Kaiser, and Steffen Reith

University of Applied Sciences, Unter den Eichen 5, 65195 Wiesbaden, Germany
alex.schoenborn@web.de

Abstract. With the goal of using the open source RISC-V ISA in real-time processing, the porting of the real-time operating system "Marron" is described in detail. The result is then extended to a hypervisor so that multiple guest systems with different timing and resource requirements can be safely separated. Since both solutions are open source, they enhance the available open source infrastructure. They thus help to reduce supply chain problems and avoid dependencies in the future.

Keywords: RISC-V · Real-Time · Operating System · Virtualization · Partitioning · Open-Source Hardware

1 Introduction

Due to the recently increased supply chain problems as well as the general dependence on individual suppliers, digital sovereignty is becoming an increasingly important issue. The open source approach is seen as a possible way out of this dilemma. While open source *software* is already well established in the form of projects like BSD, Linux, FreeRTOS, Zephyr, Apache and more, the idea of open source *hardware*, i.e. the release of complete processor designs implementing an instruction set architecture (ISA) under an open source license using open EDA tools, is relatively new.

RISC-V [9] is such an ISA which, in strong contrast e.g. to ARM, is released under a non-restrictive license allowing anyone to implement it. Consequently, numerous free implementations exist ranging from small FPGA variants [5] up to rad-hardened chips suitable for space applications [7]. On the software support side, however, things still look bleak at the moment. While open source development tools and general-purpose operating systems such as Linux are readily available, dedicated real-time operating systems suitable for safety-critical applications are still scarce.

In the PROGENITOR project [1] we strive to extend and to enrich the open source software landscape around the RISC-V open source hardware platform. To this end, this work [8] describes an effort of porting a real-time microkernel-based operating system to the architecture. As this microkernel features memory protection and resource separation, it already has the necessary ingredients to

© The Author(s), under exclusive license to Springer Nature Switzerland AG 2023
H. Unger and M. Schaible (Eds.): Real-Time 2022, LNNS 674, pp. 33–42, 2023.
https://doi.org/10.1007/978-3-031-32700-1_4

implement virtual machines. Therefore, in a second step, the development is taken further by extending the microkernel by hypervisor functionality, thus making it possible to combine subsystems with different safety requirements in a single machine without any interference.

This paper is structured as follows: Firstly, Sect. 2 gives an overview of the internal structure of the Marron real-time kernel, followed by considerations and details on the necessary steps to be taken for the porting effort. Secondly, Sect. 3 discusses virtualization in the context of RISC-V and how to integrate it into the previously ported kernel. Finally, Sect. 4 provides a brief discussion of the results obtained and Sect. 5 has some concluding remarks on the work.

2   Porting an Operating System

The effort of porting an operating system to a new architecture is often undervalued. The prevailing opinion is that it is merely a mechanical activity that offers little room for creativity and hardly any challenges worth mentioning. However, in reality, it requires a deep understanding of the underlying computer architectures, the ability to program in machine language[1] and an equally deep understanding of concurrent programming. Especially in the context of real-time operating systems, a poorly coded kernel entry or exit procedure can easily spoil a system's real-time properties.

This section describes the porting of *Marron* [11], a microkernel-based real-time operating system developed for 32-bit and 64-bit ARM computer architectures as part of the EU project AQUAS [2]. Marron supports priority-based scheduling and provides some very efficient task synchronization primitives [10]. In contrast to many common real-time executives, Marron features memory protection, thus enabling it to establish *partitions* in which user programs can execute in strict separation. This makes Marron suitable to serve as a separation kernel for MILS[2] [6] systems.

2.1   Static Configuration

One important design choice made in Marron is its static configuration: All tasks as well as their accessible resources are configured offline, at design time. Therefore, several kernel data structures such as the page tables for the memory management unit, task queues, interrupt handlers, etc. are built by a special code generator tool. As some of these data structures are architecture dependent, the code generator needs to be adapted as part of the porting effort.

Each partition configured in this way needs to be assigned its own binary payload. This is also the task of a tool that assembles the kernel and the user programs into a bootable image.

---

[1] which is not trivial, especially with RISC architectures as their microarchitecture properties need to be considered at the programming level.
[2] Multiple independent levels of security.

2.2   Kernel Structure

To facilitate porting, all kernel code is divided into three different categories:

- *Generic:* These code portions are platform and board independent. They are written in C and it should not be necessary to touch them when porting the kernel to another architecture and platform.
- *Architecture specific:* These portions are specific to the architecture (i.e. the ISA), but not to a particular hardware board. This code has to be modified (essentially: re-written) when porting to a new architecture. It can be written in C or assembly language, depending on practicability.
- *Board specific:* This part is specific to a particular hardware configuration (i.e. a board or an SoC). It typically contains a few simple device drivers and board initialization code. This code is also referred to as the *board support package* (BSP). Ideally, BSP code should not be architecture specific and is thus typically written in C. However, this kind of orthogonality can not always be achieved.

In order to keep porting effort low, architecture as well as platform specific code should be kept to a minimum. In our case, the platform specific code needs about 6.2k and the pure generic kernel code about 9.2k lines of code.

2.3   Kernel Startup

The first thing to do at system startup time is to create a valid environment for C code. At the very least, this requires setting up a stack. As this can only be done at the assembly level, the code implementing it is a natural part of the architecture specific code. Next, depending on board hardware, early initialization of some peripherals may be necessary for the board to function. The startup code is therefore divided into two parts, the first of which is in the architecture specific kernel entry code, while the second is part of the BSP. In detail, the following steps are taken during startup:

- The stack pointer register is set to point to the top of the kernel stack. It is expected that there is sufficient free memory available below this address. The stack address is provided by extra linker symbols pointing to the beginning and end of the memory area that shall serve as stack. These are defined in a linker script that is used for building the kernel binary. Since this linker script is board specific, it belongs to the BSP.
- The `bss` section (i.e. uninitialized static memory) must be cleared to 0 beforehand. This is necessary because C programs may legally expect all their static variables to be zero upon start. Again, linker symbols are provided for this.
- The `stvec` (supervisor trap vector) register is set to `riscv_stvec_entry` (address of kernel entry). This will ensure that upon any trap, the kernel will be entered via this entry point. Note that, in contrast to many other architectures, RISC-V leaves the choice of a trap vector table with entries for different kinds of traps to the programmer.

- The kernel-main function is called. This is the first piece of C code being invoked. Also, early board-specific initializations are performed by calling bsp_init().
- Various other components of the operating system, such as IRQs, threads, partitions and the scheduler are initialized. This mainly entails the allocation of free space for the creation of data structures and their subsequent initialization. Most of the procedures invoked here are in the generic part of the kernel code.
- Finally, the kernel is left by a jump into the first user level thread.

2.4   Kernel Entry and Exit

Upon entry into the kernel, the current process' registers must be saved in memory for a later resume. For this, Marron uses a platform specific substructure, the *register save area*, in the corresponding task control block (TCB) structure. To quickly locate this substructure, the kernel places its address into the sscratch status register prior to exiting into user mode. As this register is inaccessible to user mode code, it can be trusted to always contain the right value upon kernel entry. The kernel can not trust the current user stack to be valid (otherwise a corrupted user stack could easily crash the kernel). Therefore, a valid stack must be readily available upon entry into the kernel. Marron uses a single stack kernel design, i.e. all tasks use the same stack while inside the kernel. This implies that only one task at a time (per processor) can be active in the kernel, i.e. that the kernel is not preemptive. However, it is possible for the kernel to be re-entered. This can occur for example in the case of a hardware interrupt being received while executing in the kernel, or in case of a fault in the kernel code. In such cases, the kernel entry code **must not** re-set the kernel stack but instead it must continue to use the already-established kernel stack as is, effectively treating the exception as a subroutine. The general sequence of a kernel entry is:

Enter the Kernel: When an exception occurs, the processor's program counter is set to the value held in the stvec register. This register was set to riscv_stvec_entry (address of kernel entry) during initialization (see Sect. 2.3), so the kernel is entered upon exception. The previous program counter value is saved into register sepc, so the kernel can determine the location of the code causing the exception (and where to resume later). Furthermore, registers sstatus, scause and stval are loaded with further information. These will be interrogated later to determine the exact cause of the exception and an appropriate method to deal with it.

Save the Register Context: The pre-exception user space register context must be saved in the register save area of the current thread. No information must be lost, so the kernel entry procedure must not spill any registers. As explained above, the sscratch register can be trusted to contain a pointer to the current process' register save area. The CSRRW RISC-V instruction allows to atomically swap the contents of a system register (sscratch in this case) with another register (here: the stack pointer).

Switch to the Kernel Stack: The kernel has its own stack which is used during the exception handling. This pointer has just been swapped in from the sscratch register. However, if the kernel is not invoked from user space but is being re-entered, it must continue to use the already-established kernel stack. To facilitate this, the kernel makes sure that sscratch contains 0 while in the kernel. This makes it possible to place a branch on zero instruction immediately after the swap instruction. This branch then skips the re-setting of the kernel stack.

Dispatching: After these steps, the rest of the exception handling can be done by C code. Marron will map the architecture specific exceptions to generic ones: It distinguishes between interrupts, system calls and faults. This distinction is made based on values found in the sstatus, scause and stval registers.

Scheduling: After the exception has been handled, the scheduler provides a register context for execution. This can be the same thread as before or another one, e.g. if the previous thread was blocked, or in case a thread with higher priority has been unblocked.

Restore the register context: The context from the selected register save area is loaded into the actual processor registers. Compared to context saving, restoring is easy because the pointer to the save area can be overwritten at the end. This will be also the last step before exiting the kernel. Before that, CSRs have to be loaded again and some special cases distinguishing between the idle thread and a normal user thread have to be dealt with[3].

Exit the Kernel: The kernel is exited and the previously loaded context is executed. This will be achieved by using the sret instruction. Compared to a normal procedure return, sret has some additional features: sepc will be used as the target program counter. This is needed, because otherwise the RA register of the to-be-executed thread would be lost. Additionally SPIE and SPP are used to set the interrupt-enable bit and the privilege mode for the return.

2.5   Interrupt Management

Interrupts are controlled in RISC-V by mie (machine interrupt enabled) and mip (machine interrupt pending) registers. mie controls the allowed interrupts and in mip the appropriate bit is set when an interrupt is requested. If a bit is active in both registers, and unless interrupts are globally disabled through the mstatus register, a corresponding interrupt will be triggered in the CPU. Like any other exception, this will invoke the kernel entry code (see Sect. 2.4) with the scause register indicating that an interrupt is pending. The actual handling of interrupts by the kernel is quite different from that of other exceptions like faults or traps: Since it is not caused by the currently running process, it does not make sense for the kernel to look at the current process' state. However, interrupts can lead

---

[3] The idle thread is executed in system mode and thus the SPIE bit will be used to differentiate.

to process switches, therefore it is important that the current process' state is saved completely upon entry in exactly the same way as if it had entered the kernel deliberately via a system call. Furthermore, interrupts can also happen while running in the kernel, which effectively means that the kernel is being re-entered. Such cases have to be treated specifically, as was pointed out in Sect. 2.4.

The RISC-V architecture supports multiple possible interrupt sources, typically through an external, platform level interrupt controller (PLIC). However, at the time of this writing, the Marron implementation described here only supports timer interrupts as these are required for multi-tasking.

2.6 Memory Protection

Memory protection is realized with the help of page tables and the `satp` (supervisor address translation and protection) register. `satp` contains `PPN` (physical page number/root page table), `asid` and `MODE` as fields. As Marron features static memory configuration, the actual page tables are built off-line and are never modified during runtime. Upon address space switch, the `satp` register is set to point to the corresponding task's page table root. Also, individual address space IDs are assigned for each address space and are loaded into the `asid` field upon a switch. Page table entries which are cached in the Translation Lookaside Buffer (TLB) are effective only if their ASID value matches the one in the `satp` register. Thus they can be left in the cache while other address spaces are in effect, avoiding the need for a TLB flush upon an address space switch. The off-line tool for the generation of the page table (written in Python) required only a few modifications as the logical structure of two-level page tables is similar between ARM and RISC-V. Only a few bit positions and field sizes required some adjustments. However, RISC-V physical addresses are 2 bits larger than the virtual address. This created problems at the linker step for the generated page tables. The entries in the level 2 page tables could simply be adjusted, but because entries in level 1 can be pointers to level 2, this is not a viable solution. The chosen solution was to programmatically adjust the entries at startup time before the MMU gets turned on.

3   Virtual Machines in Marron

Since Marron already offers address spaces, and since the RISC-V ISA specifies a virtual machine extension, most of the ingredients for a full virtual machine are already there. Therefore, in the second part of this work, the development of a prototypical virtual machine monitor is attempted. To validate its functionality Marron is run "within itself", i.e. as the guest operating system in a virtual machine hosted by Marron. This is a simpler approach than using e.g a fully blown Linux guest.

3.1   Virtualization in RISC-V

RISC-V provides virtualization support in its privileged ISA by using exceptions. This is achieved by the following three bits in `mstatus`:
Trap Virtual Memory (TVM) will raise an illegal instruction exception on attempts to read or write `satp` or at the execution of the `SFENCE.VMA` and `SINVAL.VMA` instructions. Through this it is possible to populate the shadow page tables. Timeout Wait (TW) will trap the Wait for Interrupt (WFI) instruction and make it possible to switch to another guest, instead of staying idle in the current one. Trap SRET (TSR) will raise an illegal instruction exception on attempts to execute `SRET` while executing in S-Mode. Moreover, RISC-V provides a hypervisor extension (reached frozen state after this work) to further support virtualization. The addition of new execution modes, CSRs and additional mechanisms for traps, world-switches and page tables increase the efficiency of virtualization.

3.2   Structure

The structure of a virtual machine is usually static. Assigned devices such as UART or network cards do not change at runtime. It therefore makes sense to also map the VMs statically in the configuration of Marron. RISC-V specifies HS-Mode as an extension of S-Mode, so it would make sense to implement entry and exit to a VM (part of the VMM) in the kernel. In fact, this is necessary since access to hypervisor CSRs and the `sret` instruction are part of the privileged HS-Mode. The remaining functionality can be implemented in userspace using the microkernel approach. For memory protection in virtualization, RISC-V uses a two stage design. The page table inside the virtual machine can be ignored from the hypervisor's point of view, since the VM (as a guest) takes care of it and sets it in `vsatp`. For the second stage, the page table in `hgatp` is used. This page table belongs to the hypervisor and transforms a physical guest address into a physical supervisor address. This provides the actual memory protection between guest and host. Due to the static approach, it is a reasonable solution to integrate the generation into the code generator.

3.3   VM Entry and Exit

The entry and exit of the VM are provided via a system call with the VM context as an argument. The system call saves the host context in an additional save area so that it can be reloaded after the VM exit. The VM context is loaded into the register save area so that it is used instead of the host context when the exception handler exits. In addition, the VS-Mode and HS-Mode CSRs are set. In order to take the correct return path for the system call of the VM entry, a small adjustment is required. Before checking the `SPIE` bit the `SPV` bit in `hstatus` is checked. If `SPV` is set, this means that `sret` will return into a VM. This ensures that `sscratch` points to the correct register save area for re-entry into the kernel. The VM exit lands in the normal entry point of the kernel The exception handler must be passed the state of `hstatus` as an additional argument. This is necessary so that `SPV` can decide whether the context was virtualized before the kernel entry or not. If `SPV` is set, a separate routine will be used. First, the VM context in the register save area is saved back into the VM data structure. The previous host context is moved back into the register save area. VS-Mode and HS-Mode CSRs are also saved into the VM data structure. If the trap was an interrupt instead of an exception, the interrupt handler is called and the interrupt is also set for the VM. The handling of exceptions is located in the user space part of the VMM and is performed by the restored host context.

3.4   Interrupts

The status registers `medeleg` (Machine Exception Delegation Register) and `mideleg` (Machine Interrupt Delegation Register) allow a selection of certain exceptions and interrupts to be trapped in S-Mode instead of M-Mode. When an exception occurs, the system uses the equivalent S-Mode instead of M-Mode CSRs. This procedure is also possible between VM and hypervisor using `hedeleg` and `hideleg`. This allows exceptions to be further delegated to the VS-Mode, with a few exceptions. No ecalls from higher privileged execution modes (HS-Mode, M-Mode) can be passed on to VS-Mode. Also, of course, no G-stage pagefaults may be passed to the guest to handle. The last exception that must not be delegated is the `virtual instruction`, since the host must emulate the instruction. To inject an interrupt into the VM, the `hvip` register is used. Through this interrupts can be inserted into `hip`, which represents a "logical or" between `hvip` and the respective interrupt sources. `hip` and `hie` serve as an extension to `sip` and `sie` for virtualization.

4   Discussion

Many operating systems have already been ported to RISC-V, including Linux and several BSD derivatives [3]. Also real-time OSes like FreeRTOS, Zephyr, L4 and seL4 have already been ported, showing that there is interest in adopting this ISA. Marron is special in this comparison as – unlike most RTOSes – it supports full memory protection, isolation between separate resource partitions and virtualization, while – unlike the L4 microkernels – it uses static configuration, thus avoiding the complexity that comes with dynamic resource allocation. Also, unlike most RTOSes, Marron is explicitly designed to **not** be fully preemptive. This contradicts the frequently encountered view that a real-time kernel must be fully preemptive. Our response to this is as follows: Even a "fully preemptive" kernel inevitably has critical sections that must be executed with preemption disabled. The worst case preemption delay incurred under such a kernel is given by the longest of all critical sections. The constant enabling and disabling of preemption inside such a kernel adds to the length of the critical sections and to the duration of kernel calls in general, causing significant overhead. Plus it increases complexity. In order to judge the gain from fine-grained critical section management, one must view the longest critical section in relation to the overall duration of kernel code execution. Being a true microkernel, Marron does not have any long-running code sequences in the kernel and most of the code must be atomic anyway, so fine granular critical section management does not pay off here. A similar point has been made and demonstrated for the seL4 microkernel by Blackham et al. in [4]. Moreover, allowing preemption within the kernel requires the provision of individual kernel stacks for every single thread in the system. This causes significant memory overhead. Marron has a test application containing benchmarks that test functionalities like IRQs, exception handling, scheduling, etc. This test application has been used to verify that the port to RISC-V covers all functionalities of Marron. A subset of this test application has also been executed by the VM, thus showing the functionality of the virtualization.

5   Conclusion

This work shows a proof of concept for porting a micro kernel to RISC-V and adding virtualization. A lot of knowledge was gained through learning the necessary steps for porting, special properties, as well as the inner workings of a micro kernel design. The results were evaluated and discussed. There are still issues which could not be addressed in the scope of this work or could not be tested due to the use of an emulator (Qemu). While the test application demonstrates the functional completeness of the porting and virtualization, its performance, and especially its real-time properties could not be properly evaluated as QEMU is neither cycle-accurate nor does it emulate cache behavior. Marron is designed for embedded real-time applications, therefore it is necessary to evaluate the influence of virtualization on the real-time behavior. Important aspects here are the additional overhead caused by the virtualization layer, compliance with time guarantees and the real-time capabilities of the virtual environment itself. Moreover, several features are still missing: 64-bit support (RV64I), adding support for the platform-level interrupt controller (PLIC), multi-core support (SMP), SMP for each VM, hypervisor page tables to protect the host and virtualization support for other Marron targets.

References

1. Forschungsprojekt   Progenitor.   https://www.hs-rm.de/de/fachbereiche/design-informatik-medien/forschung/progenitor. 2021-2023. Accessed: 30 Sep 2022
2. AQUAS Consortium. Aggregated quality assurance for systems. https://aquas-project.eu/. (2019). Accessed 30 Sep 2022
3. Bamsch, B., He, W., Li, M., Waghela, S.: Porting OpenBSD to RISC-V ISA. Technical report, Department of Computer Engineering - San Jose State University (2020)
4. Blackham, B., Tang, V., Heiser, G.: To preempt or not to preempt, that is the question. In: Proceedings of the Asia-Pacific Workshop on Systems, APSYS 2012, New York, NY, USA, Association for Computing Machinery (2012)

5. Dennis, D.K., et al.: Single cycle risc-v micro architecture processor and its FPGA prototype. In: 2017 7th International Symposium on Embedded Computing and System Design (ISED), pp. 1–5 (2017)
6. Harrison, W.S., Hanebutte, N., Oman, P., Alves-Foss, J.: The mils architecture for a secure global information grid. Crosstalk: J. Def. Softw. Eng. **18**(10), 20–24 (2005)
7. Ramaswami, D.P., Heimstra, D.M., Shi, S., Li, Z., Chen, L.: Single event upset characterization of microsemi risc-v softcore cpus on polarfire mpf300t-1fcg1152e field programmable gate arrays using proton irradiation. Energy (MeV) **180**(520):65–120
8. Schönborn, A.: Portierung eines echtzeitkernels auf risc-v, sowie entwurf und prototypische implementierung eines virtualisierungskonzeptes auf dieser plattform. Master's thesis, Hochschule RheinMain (2022)
9. Waterman, A., Lee, Y., Patterson, D.A., Asanovic, K.; The RISC-V instruction set manual, volume I: user-level ISA, document version 20191213. EECS Department, University of California, Berkeley (2019)
10. Zuepke, A.: Turning futexes inside-out: efficient and deterministic user space synchronization primitives for real-time systems with IPCP. In: Völp, M. (ed), 32nd Euromicro Conference on Real-Time Systems, ECRTS 2020, 7–10 July 2020, Virtual Conference, vol. 165 of LIPIcs, pp. 11:1–11:23. Schloss Dagstuhl - Leibniz-Zentrum für Informatik (2020)
11. Zuepke, A.: The Marron Kernel. https://gitlab.com/azuepke/marron (2022). Accessed 30 Sep 2022

Predictive Preload at Fixed Preemption Points for Microcontrollers with Hard Real-Time Requirements

Philipp Jungklass[1](✉), Folkhart Grieger[1], and Mladen Berekovic[2]

[1] IAV GmbH, Rockwellstraße 12, 38518 Gifhorn, Germany
`philipp.jungklass@iav.de`
[2] University of Lübeck, Institute of Computer Engineering, Ratzeburger Allee 160, 23562 Lübeck, Germany

Abstract. Modern microcontrollers for safety-critical applications with a hard real-time requirement use scratchpad memories assigned directly to the respective processor core to increase the performance. The advantage of scratchpad memories is their deterministic and high access speed. The disadvantage, on the other hand, is the low memory capacity, which means that efficient use is essential for maximum execution speed. For this reason, this article presents a concept that predictively preloads the program scratchpads at fixed preemption points in order to effectively compensate the low memory capacity.

Keywords: Multicore processing · Memory architecture · Real-time systems

1 Motivation

Multicore microcontrollers are increasingly used in safety-critical applications [5,15]. These use local scratchpads to increase execution speed, which improves performance and reduces the number of accesses to shared memory. Due to these properties, scratchpads have a significant impact on the Worst-Case Execution Time (WCET) [4,18]. As a result of the used memory technology, scratchpads provide only a small memory capacity in contrast to flash memory, therefore the efficient usage is essential. However, it should be noted that the data scratchpads are often significantly larger in capacity than the corresponding program scratchpads, as the Table 1 exemplarily shows [7]. Therefore, this paper proposes a dynamic memory management scheme that uses scheduling based on fixed preemption points to predictively preload the program code to be executed at next into the local program scratchpad of the respective processor core.

2 Related Work

The existing methods for dynamic memory usage in real-time systems with scratchpads can basically be divided into two categories. In the first group are

© The Author(s), under exclusive license to Springer Nature Switzerland AG 2023
H. Unger and M. Schaible (Eds.): Real-Time 2022, LNNS 674, pp. 43–51, 2023.
https://doi.org/10.1007/978-3-031-32700-1_5

methods which base on different hardware adaptations. These include, for example, separate Direct Memory Access (DMA) controllers for each processor core [9], specialized Memory Management Units (MMUs) [2] or instruction-based preloading scratchpads [11]. In contrast, there are the software-based methods, which, however, often use intelligent cache management [10,19] for dynamic preloading. For example, in [13], the scratchpads are used only for static code during the caches are used for dynamic preloading.

In comparison, the method presented in this article uses a software-based approach to predictively preload the program code needed next, into the local scratchpad of the respective processor core, which effectively reduces the WCET.

3   Concept

The basic concept is that the program scratchpad is divided into two areas. One half contains the code which is currently being executed, while the program code which will be executed subsequently is preloaded in parallel in the second half. To ensure that the prediction is as precise as possible, the operating system identifies the subsequent program block during a fixed preemption point and initiates the preloading. This procedure ensures that the processor core only executes code that is already stored in the local scratchpad. In addition to the associated increase in execution speed, accesses to the shared memories in multicore systems can also be reduced.

3.1   Preload Block Identification

In the current development phase, the preload blocks are separated by manually set fixed preemption points. The advantage of this approach is that the operating system can only perform a context switch during these specific interruptions, which effectively prevents an intermediate reloading of the scratchpad. By defining the fixed preemption points, however, it must be ensured that the overhead for the context switch is as low as possible and that the memory limitation is observed as a result of the separation of the scratchpad. Due to the mentioned boundary conditions, a nearly automated setting of the fixed preemption points should be realized in future research [1,12,16]. Additional constraints, such as the use of shared resources, allowed maximum response times, and the Worst-Case Execution Path (WCEP), should also be considered [3].

3.2   Preload Procedure

For preloading into the program scratchpad, this article proposes three methods that use different features of the microcontroller.

Preload via Core. The first variant of the predictive preload does not require any special hardware features of the used microcontroller. As the Fig. 1 shows, the corresponding program code is loaded into the scratchpad by the processor core before the execution of the preload block is processed. Following this, the execution directly starts. The advantage of this variant is the simple implementation as well the fact that no memory must be reserved for the parallel preload. Furthermore, no modifications to the existing operating systems are required because only the preload block currently to be executed has to be defined, which reduces the overhead. This variant is therefore particularly suitable for low-power microcontrollers with a small hardware feature set. On the other hand, the significantly reduced performance is a disadvantage, because the preload is performed by the executing processor core itself. The Formula (1) can be used to evaluate the benefit.

Fig. 1. Preloading via executing core

$$t_{PS}(E) + t_{PS}(P) < t_F(E) \tag{1}$$

$t_{PS}(E)$ Execution time of a preload block using the program scratchpad
$t_{PS}(P)$ Preload time of a preload block
$t_F(E)$ Execution time of a preload block using the flash

Preload via Dedicated Core. In safety-critical applications with a hard real-time requirement, multicore microcontrollers are used increasingly. Due to the higher number of available processor cores, it is possible to provide a dedicated preload core in the system. This preload core handles all requests from the complete system and copies the required preload blocks to the scratchpads of the corresponding processor cores, as shown in the Fig. 2.

It is important to note that the preload core is no longer available for calculations, which reduces the available computing power. Only if the remaining processor cores can compensate this deficit by using their scratchpads is the described approach useful. The Formula (2) can be used as a basis for the evaluation.

$$\sum_{m=0}^{m_{\max}} (t_{PS}(E_{all}) + t_{PS}(P_{all})) < \sum_{n=0}^{n_{\max}} t_F(E_{all}) \tag{2}$$

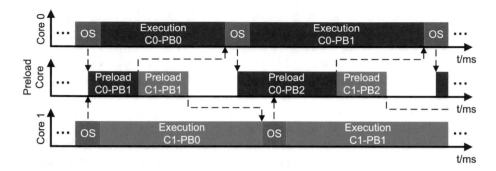

Fig. 2. Preloading via dedicated Core

$t_{PS}(E_{all})$	Execution time of all preload blocks using the program scratchpad
$t_{PS}(P_{all})$	Preload time of all preload block
$t_F(E_{all})$	Execution time of all preload block using the flash
n	Number of cores
m	Number of execution cores

$$(3)$$

Preload via DMA. A DMA controller is integrated in various microcontrollers for the fast copying of large data volumes. This is ideal for performing the job of the preload core so that the full computing time is available again for calculations, as shown in the Fig. 3.

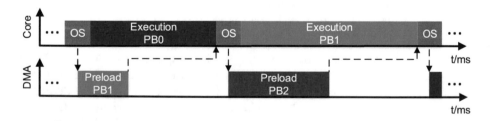

Fig. 3. Preloading via DMA

To ensure that the processor core should not wait for a transfer from the DMA controller, all preload blocks must satisfy the condition in Formula (4). This condition expresses that the maximum transfer time from the DMA controller for the next required preload block is less than the minimum execution time of the current preload block by the processor core.

$$t_{PS}(E(PB(i)_{min})) \geq t_{PS}(P_{DMA}(PB(i+1)_{max})) \tag{4}$$

$$t_{PS}(E(PB(i)_{min})) \qquad \text{Minimum execution time of preload block (i) via CPU} \atop t_{PS}(P_{DMA}(PB(i+1)_{max})) \quad \text{Maximum copy time of preload block (i+1) via DMA} \qquad (5)$$

3.3   Address Translation

After copying the program code for execution, the addresses for the functions must be modified if necessary. Due to the fact that real-time capable microcontrollers do not provide MMUs, alternative concepts must be used [5].

Table with Function Addresses. The first option is using an address table, which contains the jump addresses of all functions. The operating system can modify the addresses accordingly after a successful transfer to the local scratchpad. The advantage of this approach is the flexible use. The disadvantage of this concept of address manipulation is the resulting overhead.

Address Replacement. By using the address replacement the preload blocks are stored in the flash memory in such a way that the correct addresses for an execution from the scratchpad are already available. The advantage of this procedure is the low overhead, since no address conversion is performed. On the other hand, the low flexibility and the missing option to execute the program code directly from the flash memory, if a transfer of the program code does not succeed in time, are disadvantages.

Overlay Memory. Some microcontrollers support an overlay memory, which can be used to manipulate address accesses already in hardware. Through this feature the overhead can be reduced by an address conversion. The disadvantage of this approach is the necessary hardware support and the overhead due to the configuration of the overlay address space [6].

3.4   Asynchronic Events

In systems with a hard real-time requirement, asynchronous events, such as interrupts, are often used. Because these asynchronous events cannot be forecasted, predictive preloading is more difficult to handle. Therefore, an area of the scratchpad should always be reserved, wherin the program code for handling the interrupt is statically allocated. [8].

4   Experimental Results

To verify the functionality, the proposed method is implemented on an Infineon AUtomotive Realtime Integrated NeXt Generation Architecture (AURIX) TC399, which is mainly used for safety-critical applications with hard real-time requirements. The AURIX TC399 offers six processor cores, which use the proprietary TriCore architecture. Each processor core has a separate interface for the program code and for the data, each of these having a scratchpad and a cache. In addition to the local memories, which are directly assigned to a core, there are also two global memories in the type of flash and RAM memory [6]. The corresponding memory sizes can be taken from the Table 1.

Table 1. Infineon AURIX TC399 - Memory Dimensioning [6]

Memory	Infineon AURIX TC399
Local program scratchpad/per core	64 KB
Local program cache/per core	32 KB
Local data scratchpad/per core	240 KB
Local data cache/per core	16 KB
Global RAM	1,152 KB
Global flash	16,777,216 KB

The real-time operating system that is used is ErikaOS v3, which already supports the AURIX TC399 and is available as an open source solution. Due to the fact that ErikaOS does not natively support fixed preemption points, these are simulated by using critical sections. The Infineon Low Level Driver library in version 1.0.1.10.0 is used to control the peripherals of the AURIX. EEMBC's CoreMark is used as a benchmark for comparing the different approaches. The source code is translated using the tasking compiler for the TriCore architecture in version TC6.2r2.

For the analysis of the proposed concept, the preload via DMA controller is used, whereby the address translation is realized by using address replacement. The optimum distribution, which is shown in the Local program scratchpad X measurement series in Fig. 4, and the worst possible allocation, which is represented by the Global flash (bank 0) measurement series, serve as a reference. In the optimal case, each processor core is exclusively using its own local program scratchpad and data scratchpad, which completely prevents accesses to the global memory. This approach increases execution speed and prevents concurrent accesses completely. This ideal case represents a theoretical optimum, because in the rarest cases the complete program code of a processor core can be allocated to the corresponding scratchpad. However, it is possible in this measurement due to the low memory requirements of the CoreMark. The worst case is realized by the exclusive use of the global memories. All processor cores synchronously access the global flash and the global RAM, which drastically increases the access times and maximizes the concurrent accesses.

In the third measurement series Preload via DMA (32 bit) the procedure described in Sect. 3.2 is used. For this purpose, the CoreMark is separated into two parts, which are preloaded alternately at runtime. To realize a realistic load

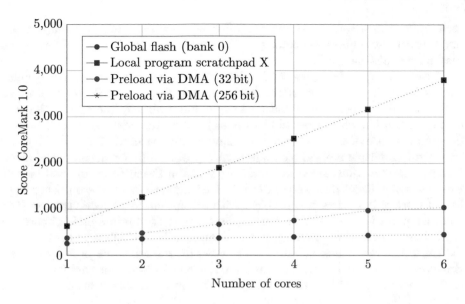

Fig. 4. Preload Performance of the Infineon AURIX TC399

scenario, a 32 KB block is transferred per preload via 32 bit transfers, which corresponds to half of the program scratchpad. As shown in the measurement, the presented method achieves about 59% of the performance of the optimal allocation in singlecore operation. The difference here is the result of the overhead of the operating system, which manages the critical section, and the configuration of the DMA transfer. As the number of processor cores increases, the performance drops to around 28% of the optimal allocation, which can be attributed to a limitation of the DMA controller. However, this result, when using six processor cores, represents an approximately 229% performance increase in contrast to the suboptimal allocation.

An optimization of the copying process is shown by the measurement series Preload via DMA (256 bit), in which the burst transfer of the DMA controller is used. This procedure can significantly increase the copying speed, which ensures identical performance compared to the optimum case.

5   Discussion

The target of this work is the more efficient use of local program scratchpads, which can be realized with the proposed concept. By predictively preloading the required program code at defined points, the low memory capacity of program scratchpads can be effectively compensated which increases execution speed and reduces the problem of concurrent accesses in multicore systems. Caches use a similar approach, but their predictability is more complex than scratchpads, which poses a problem in timing analyses [13, 17]. Compared to static allocation techniques, the advantage of dynamic approaches is that they can adapt memory contents at runtime depending on the program execution, as the presented method has also shown [8].

A disadvantage of the presented concept is the increased response time, which results from the prediction of the next preload block. In order for this to be preloaded in parallel, the operating system must already identify during the current execution which preload block is to be executed next, which can cause a delay in the case of a task change, for example. Furthermore, functions that are required in multiple preload blocks present a problem. In the current implementation, these functions are stored multiple times in the different preload blocks, which is a significant disadvantage in terms of memory consumption. Therefore, future work will address this problem with an optimized preload strategy [14]. Another point for future extensions is the support of multiple preload cores or DMA controllers. The advantage of this approach is the higher transfer performance, which can reduce potential wait cycles for the executing processor cores. However, it must be noted that this increases the focus on the problem of concurrent accesses, which is to be compensated by a memory-centric scheduling strategy [19].

References

1. Bertogna, M., Xhani, O., Marinoni, M., Esposito, F., Buttazzo G.: Optimal selection of preemption points to minimize preemption overhead. In: 2011 23rd Euromicro Conference on Real-Time Systems, pp. 217–227 (2011)
2. Chen, Z.-H., Su, A.: A hardware/software framework for instruction and data scratchpad memory allocation. ACM Trans. Archit. Code Optim. **7**, 04 (2010)
3. Dietrich C., Wägemann P., Ulbrich P., Lohmann, D.: Syswcet: whole-system response-time analysis for fixed-priority real-time systems (outstanding paper). In: 2017 IEEE Real-Time and Embedded Technology and Applications Symposium (RTAS), pp. 37–48, April 2017
4. Edwards, S.A., Lee, E.A.: The case for the precision timed (pret) machine. In: Proceedings of the 44th Annual Design Automation Conference, DAC 2007, pp. 264-265, New York, NY, USA, Association for Computing Machinery (2007)
5. Gai, P., Violante, M.: Automotive embedded software architecture in the multi-core age. In: 2016 21th IEEE European Test Symposium (ETS), pp. 1–8 (2016)
6. Infineon Technologies AG: AURIX TC3xx Target Specification V2.0.1, 07, 81726 Munich, Germany (2016)
7. Jungklass, P., Berekovic, M.: Effects of concurrent access to embedded multicore microcontrollers with hard real-time demands. In: 2018 IEEE 13th International Symposium on Industrial Embedded Systems (SIES), pp. 1–9 (2018)
8. Jungklass, P., Berekovic, M.: Memopt: automated memory distribution for multicore microcontrollers with hard real-time requirements. In: 2019 IEEE Nordic Circuits and Systems Conference (NORCAS): NORCHIP and International Symposium of System-on-Chip (SoC), pp. 1–7, October 2019
9. Kim, Y., Broman, D., Cai, J., Shrivastava, A.: Wcet-aware dynamic code management on scratchpads for software-managed multicores. Real-Time Technol. Appl. Proc. **179–188**(10), 2014 (2014)

10. Mancuso, R., Dudko, R., Betti, E., Cesati, M., Caccamo, M., Pellizzoni, R.: Real-time cache management framework for multi-core architectures. In: Proceedings of the 2013 IEEE 19th Real-Time and Embedded Technology and Applications Symposium (RTAS), RTAS 2013, pp. 45–54, Washington, DC, USA, IEEE Computer Society (2013)

11. Metzlaff, S., Guliashvili, I., Uhrig, S., Ungerer, T.: A dynamic instruction scratchpad memory for embedded processors managed by hardware. In: Berekovic, M., Fornaciari, W., Brinkschulte, U., Silvano, C. (eds.) ARCS 2011. LNCS, vol. 6566, pp. 122–134. Springer, Heidelberg (2011). https://doi.org/10.1007/978-3-642-19137-4_11

12. Peng, B., Fisher, N., Bertogna, M.: Explicit preemption placement for real-time conditional code. In: 2014 26th Euromicro Conference on Real-Time Systems, pp. 177–188 (2014)

13. Puaut, I., Pais, C.: Scratchpad memories vs locked caches in hard real-time systems: a qualitative and quantitative comparison. Institut de Recherche en Informatique et Systèmes Aléatoires - Publication Interne No **1818**, 01 (2006)

14. Rouxel, B., Skalistis, S., Derrien, S., Puaut, I.: Hiding communication delays in contention-free execution for spm-based multi-core architectures. In: Quinton, S. (ed) 31st Euromicro Conference on Real-Time Systems (ECRTS 2019), vol. 133 of Leibniz International Proceedings in Informatics (LIPIcs), pp. 25:1–25:24, Dagstuhl, Germany, Schloss Dagstuhl–Leibniz-Zentrum fuer Informatik (2019)

15. Saidi, S., Ernst, R., Uhrig, S., Theiling, H., de Dinechin, B.D.: The shift to multicores in real-time and safety-critical systems. In: Proceedings of the 10th International Conference on Hardware/Software Codesign and System Synthesis, pp. 220–229. IEEE Press (2015)

16. Suhendra, V., Mitra, T., Roychoudhury A., Chen, T.: Wcet centric data allocation to scratchpad memory. In: 26th IEEE International Real-Time Systems Symposium, p. 10 p.–232, December 2005

17. Tabish, R., et al.: A real-time scratchpad-centric OS for multi-core embedded systems. In: 2016 IEEE Real-Time and Embedded Technology and Applications Symposium (RTAS), pp. 1–11 (2016)

18. Wehmeyer, L., Marwedel, P.: Influence of memory hierarchies on predictability for time constrained embedded software. In: Design, Automation and Test in Europe, vol. 1, pp. 600–605 (2005)

19. Yao, G., Pellizzoni, R., Bak, S., Yun, H., Caccamo, M.: Global real-time memory-centric scheduling for multicore systems. IEEE Trans. Comput. **65**(9), 2739–2751 (2016)

Industrie 4.0-Compliant Digital Twins Boosting Machine Servitization

Magnus Redeker[1(✉)], Juilee Tikekar[1], Christopher Victor[2], Frank Wördehoff[2], Matthias Peveling[2], and Daniel Horenkamp[2]

[1] Fraunhofer Institute of Optronics, System Technologies and Image Exploitation, Fraunhofer IOSB, IOSB-INA, Lemgo, Germany
magnus.redeker@iosb-ina.fraunhofer.de
[2] Wöhler Brush Tech GmbH, Bad Wünnenberg, Germany

Abstract. Digital Twins are the key component for an intelligent networking of machines and processes in Industrie 4.0. Considering a machine supplier and its customers operating machines in their shop-floors, high machine availability is the cornerstone of successful business relationships. Industrie 4.0-compliant Digital Twins automatically connect the customers' machines to a servicing system of the supplier minimizing costs for machine and service integration and, through an interoperable provision of data, costs for data engineering. Machine servicing systems can be automated to the greatest possible extent. The returns on investment of service suppliers and customers increase.

Keywords: Machine Service Automation · Industrie 4.0 · Asset Administration Shell · Digital Twin Connectors · Sustainability

1 Introduction

Already in 1988 the term *servitization* was coined proclaiming that "more and more corporations throughout the world are adding value to their core corporate offerings through services" like "offering fuller market packages or "bundles" of customer-focussed combinations of goods, services, support, self-service, and knowledge" [1].

Nowadays, "Industrie 4.0 (I4.0) refers to the intelligent networking of machines and processes for industry with the help of information and communication technology" [2] causing that "value creation is increasingly shifting from production to data-based services" [3]. Value propositions for (business) customers of manufacturing companies are shorter response times of customer services, better availability of bought and operated machines and increased return on investment (ROI) [4].

Consequently, a machine supplier (seller) and a machine operator (customer) could enter into such a machine servitization business relationship contractually governing from their relation's outset

© The Author(s), under exclusive license to Springer Nature Switzerland AG 2023
H. Unger and M. Schaible (Eds.): Real-Time 2022, LNNS 674, pp. 52–61, 2023.
https://doi.org/10.1007/978-3-031-32700-1_6

- a machine's transfer of ownership from supplier to operator,
- the transmission of status data from a machine operated in the operator's shop-floor, operational and information technology (OT/IT) networks into a data platform in the supplier's IT network,
- and the supplier's servicing of an operator's machine based on its transmitted status data.

The supplier's servicing of a machine may then for example include

- the manual or automatic recommendation, booking or unmediated performance of (remote) correction or maintenance of a machine,
- the manual or automatic recommendation, booking, unmediated delivery or unmediated installation of a machine's physical spare parts or software.

Such services are based on traditional concepts like knowledge bases, best practices as well as machine learning and artificial intelligence based condition monitoring and predictive maintenance. What these concepts have in common is that they depend on historical status and solution data from a set of similar machines and current status data from a specific machine of interest.

Both, machine supplier and operator benefit from an improved machine servicing system in the short, medium and long term and strengthen their business relationship as well as their individual competitiveness. On the one hand, machine downtime can be shortened and ideally prevented, satisfying the operator's demand of a functioning and available machine. On the other hand, the machine supplier can increase revenues and reduce expenses through service processes that are automated to the greatest possible extent.

The Reference Architecture Model Industrie 4.0 (RAMI4.0) includes the concept of an I4.0 component, consisting of asset and Asset Administration Shell (AAS) [5]. An asset defines a "physical or logical object owned by or under the custodial duties of an organization, having either a perceived or actual value to the organization" [6] and the AAS concept "helps implementing digital twins for I4.0 and creating interoperability across the solutions of different suppliers" [2], where a Digital Twin (DT) defines a digital representation of an asset "sufficient to meet the requirements of a set of use cases" [6].

Machine supplier and operator consider a machine as an asset. Whereas an operator focuses on the maintenance and usage life cycle phase of a specific machine instance in which it produces goods, the supplier tries to create value throughout the entire life cycle beginning with the development of the machine type, via the type's maintenance and usage phase, via a specific machine instance's production and, finally, the instance's maintenance and usage.

A supplier's DT of a machine operated in an operator's shop-floor must integrate and provide all data across the machine's life cycle necessary to satisfy the demands of the supplier's machine servicing system. This includes historical data from the supplier's application and product life cycle management (ALM, PLM) and enterprise resource planning (ERP) systems, as well as, on the other hand, current status data from the specific machine in operation.

How I4.0-compliant DTs boost machine servitization is described in Section 2. Section 3 details exemplary machine servicing processes and, finally, Section 4 concludes this paper and gives an outlook on future enhancements.

2   Industrie 4.0-Compliant Digital Twins Boosting Machine Servitization

This section describes how I4.0-compliant DTs boost machine servitization. Considered are a machine supplier selling machines and its customers operating them in their shop-floors, OT and IT networks. Contracts govern the integration of status data from the customers' machines into the supplier's DTs as well as the supplier's machine servicing based on these integrated status data. The supplier's machine servicing depends, furthermore, on historical machine and customer data from the supplier's ERP, ALM and PLM systems.

2.1   How Industrie 4.0-Compliant Digital Twins Boost Machine Servitization

How do I4.0-compliant DTs boost machine servitization? To answer this question the following aspects must be considered.

– Systems are boosted when they become inexpensive in development, purchase, execution and recycling in comparison to the value they create.
– System integration and data engineering are particularly relevant cost factors for software systems.

Therefore, the answer to the question at hand is obvious.

– Minimization of costs for system integration and data engineering through I4.0-compliant DTs.
  • I4.0-compliant DTs integrate data into (standardized) submodels of the AAS enriching data (submodel elements) with semantic identifiers enabling humans, machines and software services to interpret the data autonomously [2,6,7]. Consequently, the effort for machine data engineering is reduced to a minimum.
  • I4.0-compliant DTs integrate all data throughout the life cycle of an asset that are necessary to implement the use cases of an organization [2,6]. ERP, ALM or PLM systems provide interface descriptions of machines, like technology specification, endpoints, security parameters, available actions, events and properties. Based thereon, adjusted DT runtimes can

automatically create, update and delete DT connectors to the interfaces of represented machines for bi-directional interaction [8,9]. Therefore, integration efforts are reduced to a minimum.
– Constant value creation through machine servicing.
  • Through servitization machine availability increases. Consequently, customers' ROI increase.
  • A to the greatest possible extent automated machine servicing process reduces servicing costs of the supplier. Frequent machine servicing create constant returns. Consequently, the supplier's ROI increases.

The bottom line is that I4.0-compliant DTs boost machine servitization because machine servicing has always been able to create value and I4.0-compliant DTs reduce servicing process costs so that the value created can significantly exceed costs – ROIs of service suppliers and customers increase.

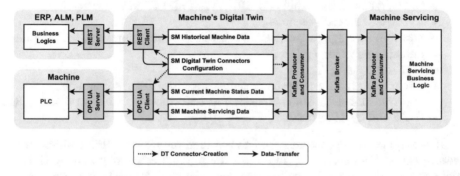

Fig. 1. A machine's Digital Twin automatically integrating the operated machine in the customer's shop-floor (current status data) and the supplier's ERP, ALM, PLM (master and historical data) and servicing systems (servicing results). Upon deployment a machine's Digital Twin creates, based on its Digital Twin Connectors Configuration submodel, HTTPS/REST-connectors to ERP, ALM and PLM as well as a Kafka connector to the machine servicing system. Based upon the integrated master data, it adapts its Digital Twin Connectors Configuration submodel and creates a further OPC UA connector to its represented machine. The Digital Twin forwards results from the servicing system to its machine for implementation and to ERP, ALM and PLM for documentation and invoicing.

2.2 Best practice

What the authors consider to be a best practice machine servicing process is described in the following. Customers' machines operated in the customers' shop-floors, IT and OT networks and, on the other hand, the supplier's machine servicing interact and exchange interoperable status data and servicing results automatically via the machines' DTs as depicted in Fig. 1.

Prerequisites and Assumptions

Prerequisites for the best practice are as follows.

- An always up-to-date list of machines to be serviced is retrievable via the API of the supplier's ERP system.
- For each of these machines
  - historical servicing relevant machine data is retrievable via the APIs of the supplier's ERP, ALM or PLM systems,
  - a remote connection via the Internet can be established in order to integrate current machine status data and transfer back servicing instructions,
  - connection establishment parameters are retrievable via the APIs of the supplier's ERP, ALM or PLM systems.
- DT runtimes are able to create, update and remove DT connectors during DT execution [8,9]
  - to the APIs of the supplier's ERP, ALM and PLM systems,
  - to the customers' machines,
  - and to a message broker for asynchronous interaction with the supplier's machine servicing system,
  in order to
  - integrate servicing relevant machine data into DTs' submodels,
  - and transmit data from submodels to machines, machine servicing, ERP, ALM and PLM systems.

It is assumed, that customers' machines each provide an OPC Unified Architecture (OPC UA) server. To minimize the manual effort required by the customers to enable remote access to these OPC UA servers via the Internet, reverse connection establishment through the servers is recommended. Unlike for the usual client-initiated OPC UA connection, a firewall in the customer's network only has to allow outgoing Transmission Control Protocol (TCP) connections to the Internet for OPC UA's reverse connection mechanism [10,11], which is usually the case in default firewall settings. Before delivery of a machine to a customer, the required connection parameters (endpoints, certificates, etc.) can be integrated into the OPC UA server as well as into the supplier's ALM or PLM systems.

Furthermore, it is assumed, that ERP, ALM and PLM systems provide Hypertext Transfer Protocol Secure/Representational State Transfer (HTTPS/REST)-interfaces, and that DTs and the machine servicing system interact asynchronously via Apache Kafka for higher performance and avoidance of blocking and crashing of the DTs and machine servicing system.

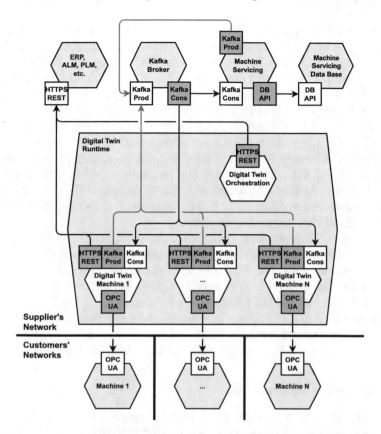

Fig. 2. A machine servicing process automated to the greatest possible extent. A Digital Twin Orchestration service periodically retrieves from the supplier's ERP, ALM and PLM systems those machines in the field that shall be serviced. New machines' Digital Twins are created and equipped with required connectors and, conversely, a Digital Twin is removed including its connectors when a machine shall no longer be serviced. White rectangles represent inbound connectors, gray outbound connectors. The machines' Digital Twins automatically integrate the operated machines in the customers' shop-floors and the supplier's ERP, ALM, PLM and servicing systems.

Machine Servicing Process

Apart from optional manual interventions in the selection of suitable service methods for a machine, the machine servicing process is automated to the greatest possible extent as illustrated in Fig. 2.

– A DT Orchestration service integrates periodically from ERP an up-to-date list of those customers' machines to be serviced, creates and orchestrates their corresponding DTs and equips each of them with connectors to ERP, ALM, PLM (HTTPS/REST) and the machine servicing (Apache Kafka).

- A machine's DT regularly integrates historical service relevant data of its represented machine from ERP, ALM and PLM (HTTPS/REST) and forwards updates asynchronously to the machine servicing (Apache Kafka).
- A machine's DT configures and creates, based on ERP, ALM and PLM data, an OPC UA client and makes it available exclusively for reverse connection establishment (OPC UA) by the represented machine's OPC UA server.
- A machine's OPC UA server establishes a TCP connection to the DT's OPC UA client, which in return establishes the OPC UA connection. If the establishing is unsuccessful or if an OPC UA connection breaks down, the machine's OPC UA server restarts the establishing again after a reasonable period of time, so that the machine's shop-floor availability is not affected.
- Upon establishment of an OPC UA connection, a DT subscribes the current status data of its represented machine.
- A machine's DT integrates status changes (OPC UA) and forwards them asynchronously to the machine servicing system (Apache Kafka).
- Upon reception of current machine status data, the machine servicing system decides for the best individual service option for the machine affected, see Section 3, and returns the result to the machine's DT (Apache Kafka).
- Upon reception of a machine servicing result, a DT forwards it
  - to its represented machine (OPC UA) for automatic implementation,
  - to the supplier's ERP, ALM and PLM for documentation and invoicing.

2.3 Alternative Approaches

A promising alternative for OPC UA's reverse connect mechanism is OPC UA's PubSub communication model decoupling OPC UA server and client by use of a middleware like an MQTT broker available in the Internet. Like for reverse connect, data can be exchanged securely and the firewall in the customer's network only has to allow outgoing TCP connections to the Internet. This is also true for plain MQTT or Apache Kafka communication with OPC UA-MQTT or OPC UA-Apache Kafka bidirectional gateways implemented in the machines and an MQTT or Apache broker available in the Internet. Such gateways could publish machine status data from OPC UA servers and receive back machine servicing instructions integrating them into OPC UA servers.

In machine servicing scenarios that require only little machine data, it could take less effort in the short term to manually engineer data and connections than to create templates for machines' DTs and to setup a DT runtime and orchestration service. However, in the medium and long term, because of the reasons stated above, the DT method would pay off.

3   Exemplary Machine Servicing

This section presents two exemplary machine servicing processes with focus on the servicing system.

Customer	Machine-ID	Machine-Type	Error-Code	Correction-Priority	Solution-Proposals
Redeker	32657	Racer	04461	3	Link
Tikekar	33102	Breaker	10820	2	Link
Horenkamp	33181	Finisher	01555	1	Link

Fig. 3. A service dashboard calling the attention of the machine supplier's staff to the customers' active machine errors. Errors are ordered according to their priority. Linked specific dashboards provide customer, machine and error details as well as solution proposals.

3.1 Error Correction Process

A machine operated on a customer's shop-floor detects system errors. These errors can be so serious that an emergency stop of the machine is carried out immediately or with a high probability must be carried out in the near future. Restoring machine availability as quickly as possible respectively preventing an emergency stop of the machine are in the most urgent interest of both the customer and the machine supplier.

As soon as an error is detected or as soon as an error status changes from active to inactive, the supplier's DT integrates the interoperable error information provided by the machine and forwards it to the machine servicing system. Based on a received error code and an internal mapping table the priority of error correction is determined.

A service dashboard, as depicted in Fig. 3, calls the attention of the supplier's staff to the active machine errors ordering them according to their correction priorities. The dashboard proposes solutions based on a knowledge data base mapping error codes to successful solutions from the past. In line with the customers' individual service agreements, the supplier's staff correct the machine errors.

In ideal circumstances, an error is correctable automatically. Staff select the solution in a machine- and error-specific dashboard, that informs the affected machine's DT. The DT forwards the solution to its very machine, which implements it automatically and restores its availability.

Alternative solution approaches are as follows.

- Staff access an affected machine remotely and correct errors manually.
- Dashboard transmits instructions to customer by mail.
- Staff transmits instructions to customer by mail or telephone.
- Spare parts are supplied.
- On-site service is scheduled.

Finally, the affected DT provides the supplier's ERP, ALM and PLM systems with the required information for documentation and invoice preparation. Furthermore, the servicing system stores error code and successful solution to the knowledge data base.

Unnecessary time losses, for example due to delayed or missing error transmissions from customer to supplier, can be avoided in all mentioned approaches. Machine emergency stops can be avoided ideally, or machine availability can be restored as quickly as possible.

3.2   Condition Monitoring and Predictive Maintenance

In addition to error information, DTs integrate and forward to the machine servicing system also process and status data from their represented machines. The servicing system, based on these very data, tensor decomposition and machine learning methods, determines machines' conditions and proactively suggests predictive maintenance to the supplier's staff in a maintenance dashboard. Predictive maintenance based on condition monitoring ideally prevents machine errors maximizing machine availability.

4   Conclusion

Industrie 4.0-compliant Digital Twins minimize costs for machine and service integration as well as for data engineering through an interoperable provision of data and through an automatic integration of a supplier's servicing system and the machines operated in the customers' shop-floors. The customers' returns on investment increase through a servicing-related increase of machine availability, and a supplier's return on investment increases through frequent servicing automated to the greatest possible extent. On the bottom line, through machines' Digital Twins the value created by servicing significantly exceeds its costs – machine servitization is boosted.

Concretely, the supplier's Digital Twins of the customers' machines automatically connect

- to the supplier's product and application life cycle management (ALM, PLM) and enterprise resource planning (ERP) systems
  - to integrate historical machine data
  - and to provide machine servicing results back to these systems for documentation and invoice preparation,
- to the customers' machines
  - to integrate up-to-date status data
  - and to provide machine-specific servicing instructions back to the machines,
- and, finally, to the supplier's machine servicing system
  - to provide historical and up-to-date status data of the represented machines to the system
  - and to receive back machine-specific servicing results from the system.

The bottom line is that the supplier's Digital Twins automatically close the integration gap and act as an intermediary between the customers' machines and the supplier's machine servicing system. Exemplary described machine servicing processes implementing condition monitoring, predictive maintenance and error correction pointed out how these Digital Twins automate machine servicing processes to the greatest possible extent.

Some prerequisites must be fulfilled. Communication endpoints must be known in advance for the automatic configuration of Digital Twin connections, and, furthermore, firewalls of supplier and customers must allow cross-company communication. Likely, Gaia-X Federation Services will further facilitate the cross-company exchange of machine data and servicing instructions and documentation.

Acknowledgments. The "Rahmenprojekt Technologietransfer" is funded by the Ministry of Economic Affairs, Innovation, Digitalisation and Energy (MWIDE) of the State of North Rhine-Westphalia within the Leading-Edge Cluster "Intelligent Technical Systems OstWestfalenLippe" (it's OWL) and managed by the Project Management Agency Jülich (PtJ). The authors are responsible for this publication's content.

References

1. Vandermerwe, S., Rada, J.: Servitization of business: Adding value by adding services. Eur. Manage. J. **6**(4), 314–324 (1988). https://doi.org/10.1016/0263-2373(88)90033-3
2. Asset Administration Shell - Reading Guide. https://www.plattform-i40.de/IP/Redaktion/EN/Downloads/Publikation/AAS-ReadingGuide202201.html
3. Digital business models for Industrie 4.0. https://www.plattform-i40.de/IP/Redaktion/EN/Downloads/Publikation/Digital-business-models.html
4. Huikkola, T., Kohtamäki, M.: Business Models in Servitization. In: Kohtamäki, M., Baines, T., Rabetino, R., Bigdeli, A.Z. (eds.) Practices and Tools for Servitization, pp. 61–81. Springer, Cham (2018). https://doi.org/10.1007/978-3-319-76517-4_4
5. DIN SPEC 91345: Reference Architecture Model Industrie 4.0 (RAMI4.0) (2016). https://doi.org/10.31030/2436156
6. Details of the Asset Administration Shell, Part 1, V3.0RC02. https://www.plattform-i40.de/IP/Redaktion/EN/Downloads/Publikation/Details_of_the_Asset_Administration_Shell_Part1_V3.html
7. 2030 Vision for Industrie 4.0 - Shaping Digital Ecosystems Globally. https://www.plattform-i40.de/PI40/Navigation/EN/Industrie40/Vision/vision.html
8. Redeker, M., Weskamp, J.N., Rössl, B., Pethig, F.: A Digital Twin Platform for Industrie 4.0. In: Curry, E., Scerri, S., Tuikka, T. (eds.) Data Spaces, Springer, Cham (2022). https://doi.org/10.1007/978-3-030-98636-0_9
9. Redeker, M., Tikekar, J.: Connecting Industrie 4.0 Digital Twins during Execution to Other Components' Interfaces. In: 27th IEEE ETFA (2022). https://doi.org/10.1109/ETFA52439.2022.9921532
10. Sichere unternehmensübergreifende Kommunikation mit OPC UA. https://www.plattform-i40.de/IP/Redaktion/DE/Downloads/Publikation/sichere-kommunikation-opc-ua.html
11. OPC Unified Architecture, Part 6, 1.04, Section 7.1.3: Establishing a connection. https://opcfoundation.org/developer-tools/specifications-unified-architecture/part-7-profiles/

A Vibrotactile Assistance System for Factory Workers for Accident Detection and Prevention in the Logistics Sector

Kevin Blümel[✉] and Michael Kuhl

Hochschule Mittweida - University of Applied Sciences, Professorship System
Electronics, Technikumplatz 17, 09648 Mittweida, Germany
`bluemel@hs-mittweida.de`

Abstract. For accident prevention in the intralogistics sector, a vibrotactile warning system in the form of an electronic-integrated safety vest
for the factory worker is presented. A multisensory approach is adopted
for better recognition of the detected area. The sensor models used and
their interaction in hazard detection are considered. The focus will be on
parallel real-time processing in data acquisition, object detection, direction evaluation and hazard detection, as well as an output of the warning
to the worker. A prototypical design and possible improvements of the
system are summarized in the final section.

Keywords: Sensor technology · assistance systems · real-time signal
processing · transportation · intralogistics

1 Introduction

In the industrial environment, modernization or digitalization is taking place on
an ongoing basis in order to relieve the workforce and make production more
effective [1]. The decisive aids are, for example, autonomously driving forklifts
and other computer controlled machines in the area of intralogistics. In addition to the regulated autonomous processes, the factory worker in a production
hall moves in a safety zone where his health must come first. To ensure safety,
automated safety routines must detect potential hazards at an early stage and
respond effectively. Due to the external influences in a production hall, e.g. noise
during metalworking, heat from blast furnaces, or glaring light from welding
work, several of the worker's senses, in this case visual, auditory perception and
thermoreception, are flooded and cannot be used to clearly perceive hazards or
information. The human haptic sense remains largely unaffected in the described
environment and can be used for vibrotactile information transfer.

H. Unger and M. Schaible (Eds.): Real-Time 2022, LNNS 674, pp. 62–71, 2023.
https://doi.org/10.1007/978-3-031-32700-1_7

2   Defining Challenge

According to [2], in the year 2020 in the German industry 43,1% of all accidents with floor conveyors were caused by forklift man collisions. A vibrotactile mobile warning system is designed to alert factory workers to hazards outside their field of vision, thus preventing accidents. For the realization of such a system, it is first necessary to reliably capture the area sensorially. Subsequently, detection of objects must take place and their danger to the worker must be determined. If an increased risk is detected due to an object in the detection area, a directional warning must be issued immediately to the worker so that he can react in time. To consider the maximum latency of the system, key timing components have been shown in Fig. 1.

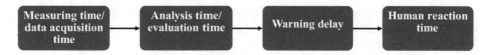

Fig. 1. Key timing components - relevant delays in an assistance system

Assuming a maximum speed of 25 km/h (≈ 7 m/s) for a potentially dangerous object and a maximum length of 5 m of the detection cone results in a maximum delay of 714 ms for the vibrotactile assistance system as shown in 1.

$$t = \frac{s}{v} = \frac{5\,[m]}{7\,\left[\frac{m}{s}\right]} = 714\,ms \tag{1}$$

where s is the maximum distance of a detectable object; v is the speed of the object.

As it can be seen, the component "Human reaction time" is not part of the electrical system and depends on the workers' condition. A prolonged reaction time of 600 ms is assumed because the worker is focused on the work and is not aware of an abrupt warning. 114 ms are remaining for the timing of the other components which are part of the electrical warning system and will be examined in the next sections.

3   State of the Art Ultrasound, Radar, Time-of-flight Camera

It is plausible that hazardous bodies (vehicles, containers to be transported, etc.) can consist of different materials that cannot necessarily be detected by a single sensor system alone due to their physical properties. Table 1 presents examples of materials and surfaces and their detection possibilities with three different sensor principles and shows that reliable detection of objects can only be made by using several sensor principles.

Table 1. Material detectability with Ultrasound, Radar and Tof camera

Materials	Ultrasound	Radar	Time-of-flight camera
Metal:	detectable	detectable	1
Wood:	detectable	2	detectable
Foam:	bad detectable	2	detectable
transparent materials: (e.g. Glass)	detectable	detectable	not detectable
Carton:	detectable	detectable	detectable

[a] depending on reflectance
[b] depending on material moisture/ conductivity

All three sensor principles work according to the "time-of-flight principle" [3]. A pulse is emitted by the sensor. This pulse hits a surface and the signal is reflected, transmitted or absorbed in different proportions. The proportions depend on the type of emitted pulse, material, surface properties, angle and other factors. The decisive factor here is the reflected signal that returns to the sensor. By measuring the time, frequency and amplitude from the start and emission of the transmission pulse to the re-reception of a partial pulse, conclusions can be drawn about the distance and position of objects in the detection range. Depending on the type of "pulse" and the related way of propagation, absorption, transmission and reflection, different materials can be detected better or worse, as mentioned above. For example, radar detects electrically conductive materials very well, but dry and insulating materials very poorly. Ultrasound, on the other hand, detects all smooth and solid bodies very well, but only poorly objects with foam-like surface structure.

The sensors of the first prototype used for the assistance system are presented below as examples:

1. **Radar** "Position2Go development kit" from the manufacturer Infineon. This sensor works with a frequency of 24 GHz. Up to five objects with direction information (angle, distance, speed) can be recognised per measurement. For data transfer USB is used.
2. **Ultrasound** "SRF10" from the manufacturer Devantech. This sensor works with a frequency of 40 kHz. Thereby the distance to the shortest reflected point can be recognised. IIC, a serial two-wire bus, is used for data transfer.
3. **Time-of-flight camera** "Terabee 3Dcam" from the manufacturer terabee. This sensor works in the infrared range. Thereby 80 × 60 voxels can be recognised per measurement, where each voxel contains a distance value in Millimeters. For data transfer USB is used.

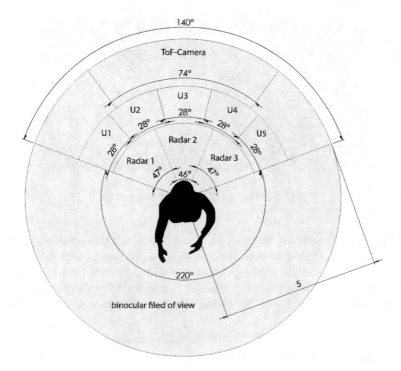

Fig. 2. Arrangement of sensors, U - Ultrasound

Decisive for the arrangement of the sensors are their respective aperture angles. The human binocular field of view is 220°. Accordingly, 140° cannot be seen without a turning movement of the head. This area is to be covered by the warning system with the sensors used. Figure 2 shows the arrangement of the sensors. For the multisensory approach, three radar sensors, five ultrasound sensors and a time-of-flight camera are used, which is visible in the overlaps of the detection areas of the sensors.

4   Concept

Figure 3 shows the concept of the warning system. The system is divided into three sections: the sensor system with ultrasound sensors, radar sensors and tof camera to detect hazards near the plant worker, the actuator system to alert the plant worker to detected hazards and the data analysis system itself.

The actuator system consists of five vibration motors, which can be controlled individually. By arranging the motors at equal distances in a row or horizontally on the back of the worker, the directional danger warning has to take place via the vibration of individual motors. The actuator system is controlled by a microcontroller and communicates with the sensor system via a serial interface. The time of sending a vibration command until the vibration is 5 ms and is

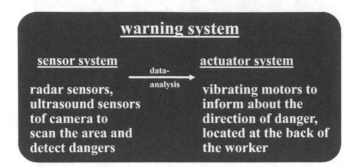

Fig. 3. Concept of the warning system

more than 80% due to the inertia of the motor. The focus of this article is on the sensor system, since this is where the bulk of the delays occur. The tasks of the sensor system include reading the sensor data, evaluating the sensor data, performing a hazard analysis, and sending commands for directional vibration to the actuator system. In the following, these tasks are underpinned by facts and figures.

4.1 Reading in the Sensor Data

The last Sect. 3 presented the used sensors as well as their number, arrangement and type of measurement data output. In order to be able to estimate the computing power for the respective sensor principle, further characteristic data of the sensor principles used must be presented. Table 2 shows the measurement time, the data packet size and the total time per sensor principle. The measurement time for each sensor was determined by measuring the time "Send start command to start a measurement" to "Receive measurement data and store it". Sensors of the same sensor principle cannot measure simultaneously because emitted pulses of the same frequency at the beginning of each measurement would falsify the measurement results among each other. The total time per sensor principle is therefore calculated by multiplying the measurement time by the number of sensors used.

As you can see the defined 114 ms for the warning system will exceed by ultrasound and radar if all sensors measure before next program part starts. The

Table 2. Data and timings of the sensors

Sensor:	Ultrasound	Radar	Time-of-flight camera
data packet size:	2 bytes	145 bytes	4850 bytes
measurement time:	38 ms	57 ms	33 ms
number of sensors:	5	3	1
total time per sensorgroup:	190 ms	171 ms	33 ms

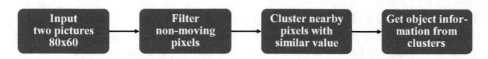

Fig. 4. Tof camera algorithm for getting objects from raw data

idea of having all sensors measured and then performing the evaluation must be discarded. If the measuring process of one sensor is complete, the evaluation is performed, after which it is the turn of the next sensor. The result of each sinlge sensor will be taken into account later on Subsect. 4.3.

4.2 Evaluating the Sensor Data

The output values of a measurement per sensor model were presented in Sect. 3. After the sensor readings have been read in, further processing or evaluation of the sensor data must take place so that a hazard analysis can be performed. The tof camera supplies only raw data in the form of a voxel array. Objects must now be extracted from this raw data. Since only moving objects are relevant for the warning system, the image is filtered first. By using two consecutive images, it is now possible to distinguish between "moving voxels" and "non-moving voxels", whereby the "non-moving voxels" are not considered further. A cluster algorithm now groups contiguous "moving voxels" with only small distance differences. The mean point as well as the mean distance of each cluster is determined. With this information and the characteristic values of the tof camera, an object can be generated from each cluster with distance and location. Figure 4 illustrates this algorithm.

It is easy to understand that this algorithm iterates several times over the relevant voxels, initially $80 \times 60 = 4800$ voxels. Measurement with the used "NUC11 Mini PC" (processor: i7-1185g7 4 GHz, ram: 2×8 gb ddr4 3200 Mhz) results in 32 ms for one run. Whenever a new measurement is available, the old image is used as a comparison image to the newly measured image in order to keep the difference between the measurements as small as possible and thus also to make small voxel value changes visible, i.e. also objects with little movement. By comparing and matching calculated objects, the velocity can also be calculated over several measurements. Thus, after the evaluation of a measurement of the tof camera, a list of moving objects with distance, angle and speed is available.

The radar sensor provides a list of up to five objects as the measured value. The object properties include distance, angle and speed in each case. An evaluation does not have to be carried out for this sensor model, since the relevant information for a hazard analysis is already available.

The ultrasound sensor provides only one distance value per measurement to the nearest reflection point. With this measurement, it is not possible to generate objects with position information. For this reason, a reduced opening angle was selected for this ultrasound sensor. With the help of the increased number of sensors (count = 5) it is possible to determine the nearest reflection point in five individual sectors independently of each other. Through two successive measurements, a velocity can now be determined in the event of a change in distance. If an object moves in the direction of the sensor in the respective sector, this is detected. The distance can be determined exactly, the angle can only be estimated and corresponds to the degrees of the sensor aperture angle as already shown in Fig. 2.

4.3   The Hazard Analysis

In order to detect a hazard and to be able to output it directionally, a coordinate transformation from the respective sensor coordinate system to the coordinate system of the warning system must take place, since the sensors are installed at an inclination to be able to scan the respective detection area. [4] is describing rotation and translation. The sensorsystem is using the translation of x and y and the rotation of x. The z is the axis which is looking from the worker in the area and the x is the axis which is looking from the back to the head of the worker.

Subsequently, the transformed object data of the radar sensors, ultrasound sensors and tof camera can be used to analyze the danger. The danger exists when a detected object approaches the warning system at a speed of more than 5 km/h. There is a subdivision into low danger at up to 5 km/h and high danger at more than 10 km/h. In the event of an existing hazard, the vibration motors of the actuator system are triggered in a direction-oriented manner depending on the object angle. The intensity of the vibration is 50% for low danger and 100% for high danger. Figure 5 The hazard analysis part takes little computing power and lasts an average of 3 ms.

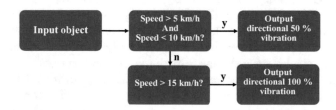

Fig. 5. hazard algorithm for generating vibration output of the aktuator system

5   Real-Time Analysis

If all the processes described were to run in a polling loop, the maximum time specified in Sect. 2 would be exceeded. To ensure that all subprocesses run smoothly in real-time, a multithread system is implemented to allow several

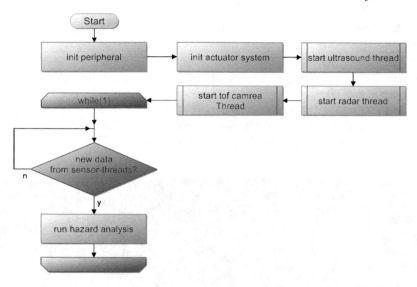

Fig. 6. program flow chart of the main task of the warning system

processes to run simultaneously. In the following program flow chart Fig. 6 the structure of the warning system becomes visible. After the start of the program and initialization, further threads are started in the main task in order to be able to carry out measurements simultaneously with radar sensors, ultrasound sensors and the tof camera. The sensor threads are starting the measurement, reading in the sensor raw data, running the evaluation of the raw data and saving the data in a loop. The main thread is waiting for new object data and is running the hazard analysis in a loop.

The timings of the warning system components were shown in the last sections. By remembering the timing of the actuator system (5 ms) and adding for example the timings for the tof camera of the sensor system (33 ms + 32 ms + 3 ms = 68 ms) we get in total 73 ms what is within the time limit of the warning system (114 ms) determined in Sect. 2.

6 Implementation and Verification

To enable the factory worker to wear the warning system during work without being hindered by it, it is mounted on a back protector in the first prototypical setup. It is important here that none of the sensors are covered and that the interfaces are freely accessible for debugging. Figure 7 shows on the left side the prototype of the warning system.

A moving pallet stack approaching the worker in the detection area of the warning system is to serve as the test object. As shown in Figure 7 on the right side, the live image from the tof camera is tracked via an external computing unit. At this moment, the worker is concentrating on a sorting task, her auditory sense is suppressed by hearing protection and her visual sense cannot perceive the pallet stack. As soon as vibration is triggered due to the approaching object, the worker raises her hand as an acceptance signal. The test is successful in several positions and environments.

To ensure the necessary real-time capability, on average 20% of the processor performance and 30 mb of the process memory were used. The present constellation has a high energy consumption. In detail this means a power consumption of 2,3 A at 19 V, which would require a battery capacity of 20 Ah for an 8h operation. This is too high for a practical application.

Fig. 7. left - prototype of the warning system, right - evaluation environment

7 Conclusion

A vibrotactile assistance system integrated on the back of a factory worker was described and conceptualized in terms of timings and data rates, as well as considerations for a multithreaded application.

Due to the required high computing power of the system, a large computing unit and thus also large accumulators are necessary. As a result, the system described increases in weight and size and would restrict the worker in everyday life. It was made visible that a large part of the computing power is required by the tof camera for object computation. Camera data cannot be excluded due to the multisensory approach. The next development is to transfer the computationally intensive tof camera algorithms to higher-level computer architectures via cloud connections. The mobile applications only need a microcontroller instead of a mini pc. In this case, however, it is imperative to ensure fast and highly stable, redundant data transmission.

A system without a tof camera could be integrated into a safety vest without a server and without a large computing unit. In this case, however, it would be necessary to be able to determine exact object positions with the help of the radar sensors and also ultrasound sensors. In [5], therefore, an ultrasound system is described that enables accurate object detections with ultrasound sensors based on triangulation. It is important to use ultrasound sensors that record the amplitude over time and output its measurement points. With this raw data, object distances and angles can be detected. Radar and ultrasound sensors could provide this system with measurements for object detection. A microcontroller would be sufficient for the computing power of such a system, saving a lot of space and weight. The system described in this article as a multithreaded application can also be implemented on the microcontroller without a camera application. However, it should be noted that the threads cannot run in true parallel as long as the CPU has only one core. An automaton would have to be designed for an optimized real-time application.

References

1. Blümel, K., Tagliaferri, F., Kuhl, M.: Algorithm for calculating distance and sensor-object angle from raw data of ultra-low power, long-range ultrasonic time-of-flight range sensors. In: 16th CIRP Conference on Intelligent Computation in Manufacturing Engineering, ScienceDirect (2022)
2. Deutsche Gesetzliche Unfallversicherung (DGUV), Statistik Arbeitsunfallgeschehen 2020 (2021). https://publikationen.dguv.de/widgets/pdf/download/article/4271. Berlin (S.82)
3. Marioli, D., Narduzzi, C., Offelli, C., Petri, D., Sardini, E., Taroni, A.: Digital time-of-flight measurement for ultrasonic sensors. IEEE Trans. Instrum. Measur. **41**(1), 93–97 (1992)
4. Reinhart, G., et al.: Der Mensch in der Produktion von Morgen, Carl Hanser Verlag GmbH & Co. KG., München (2017). (S.51-88)
5. Stark, G.: Robotik mit Matlab, Carl Hanser Verlag, München (2022). (S.54-62)

MBSE for SMEs with Domain-Specific Safety Analyses and Loose Tool Coupling

Nick Berezowski[✉] and Markus Haid

Hochschule Darmstadt, CCASS, Birkenweg 8, 64293 Darmstadt, Germany
berezowski@ccass.de

Abstract. Nowadays, systems such as motor vehicles, medical devices, or industrial machines are designed according to functional safety requirements. The main criterion is the safety integrity level (SIL), which is mainly based on failure probabilities. A development approach based on models, also called Model-Based Systems Engineering (MBSE), is increasingly coming into focus for a highly regulated development process. Probabilistic safety and reliability analyses for calculating failure probabilities and applying these in MBSE are currently only integrated to a limited extent. Merging partial development processes increases confidence and couples the design process with all stakeholders. This paper presents a general method that develops domain-specific extensions based on specifications from Object Management Group (OMG) regarding risk analysis and assessment modeling for more practical applicability of MBSE for functional safety.

Keywords: Safety · Domain-Specific Analysis · SysML · Loose Tool Coupling

1 Introduction

The standard IEC 61508 describes the basis for functional safety system development. It sets requirements for the entire safety life cycle, from the conception phase through system development to the safety-related products' decommission [1]. For the Model-Based Systems Engineering (MBSE) approach published by the International Council on Systems Engineering (INCOSE) several modeling languages have been developed, including Architecture Analysis and Design Language (AADL), Lifecycle Modeling Language (LML), Unified Modeling Language (UML), and Systems Modeling Language (SysML). SysML is a profile defined at the meta-model level that further specializes UML for technical system developments [2]. SysML is intended to provide more in-depth support for specification, design, analysis, verification, and validation [2]. According to D'Ambrosio et al. [3], SysML can be understood as the preferred language for systems modeling today. Open source development tools like Papyrus from Eclipse fully support the current version of SysML.

H. Unger and M. Schaible (Eds.): Real-Time 2022, LNNS 674, pp. 72–81, 2023.
https://doi.org/10.1007/978-3-031-32700-1_8

The core for evaluating the functional safety of a design are probabilistic safety and reliability analysis methods such as failure mode and effect analysis (FMEA), fault tree analysis (FTA), and other more complex and dynamic procedures that involve time [1,4,5]. In domain-specific standards, which represent extensions of the IEC 61508, hazard classifications vary, as do the obligations to use special methodological procedures [1]. FTA is a static analysis method in which an undesirable system state is analyzed using Boolean gates, like AND and OR, that combine a series of intermediate and basic events. The basic idea is to analyze failure behavior to determine and demonstrate the value of remedial actions and determine failure rates for safety-related system failures [1]. Based on the system definition, the structure of the fault tree can be derived by identifying the events that can lead to any undesirable system behavior [6]. The final step of quantitative assessment calculates the probability of a resulting system failure based on probabilities of the basic events [6]. Tools exist, such as CAFTA from the Electric Power Research Institute, EMFTA in Eclipse, or the ALD Fault Tree Analyzer. However, the static FTA model cannot represent behavior such as the sequence of fault events [7]. Therefore, static fault trees often cannot describe the behavior in enough detail for practical systems due to their too simple assumptions and lack of common dependability patterns [5]. FMEA is a technique used to systematically determine and evaluate the consequences of individual failures of various types [1]. FMEAs are fundamental analyses that can be used in almost all domains.

There is a clear advantage to extend actual research for domain-specific needs concerning MBSE. In this way, improvements in the quality and efficiency of development processes in companies with limited resources, established procedures, and established analysis tool environments can be made. A linkage between system and safety models improves communication and consistency between stakeholders. The OMG Risk Analysis & Assessment Modeling Language Specification (RAAML) [6] addresses these problems and wants to offer a possible solution. Their 1.0 beta version was officially released on Sept. 15, 2021. RAAML extends the SysML language capabilities into the safety engineering domain [6] using profile extensions to build a concept incorporating domain-specific safety and reliability analyses into safety-related system development. It consists of a set of profiles and libraries organized in a networked package structure, in which there are three main packages called "Core", "General", and "Methods" [6]. "Core" specifies basic notions and relationships for safety and reliability and "General" extends the "Core" for the method- or domain-specific packages described in "Methods". Each package consists of a profile and a library. The packages and profiles can thus be used and extended for further domain-independent, domain-related or project-related analyses. Each event in the FTA profile contains a probability and a priority, whereby the probability can only be adjusted in BasicEvents, ConditionalEvents, UndevelopedEvents, and DormantEvents. The probabilities of the remaining event types are derived and calculated by performing the analysis in the development environment [6]. A probability in FTATree determines the failure probability of the whole modeled system [6].

If system and analysis models are developed in different modeling languages, this can be called loose model coupling and if all existing aspects are consistently modeled in an integrated model, this can be understood as tight model coupling. If tools are addressed externally over translators, it is called loose tool coupling. Current development processes often only rely on loose model coupling [4,8, 9]. Loose model coupling can have costly consequences if stakeholders are not synchronizing system model changes directly with the analysis models. If analysis tools are addressed externally via translators, this is called loose tool coupling, otherwise, i.e. when calculations are made in the modeling environment itself, this is called tight tool coupling. The OMG approach applies to tight model coupling and a tight tool coupling.

The paper represents an extension and a concretization of MBSE for safety-critical areas. The proposed approach described in Sect. 2 is based on the RAAML approach as well as research that already addresses individual safety and reliability analysis methods. To demonstrate the approach, dynamic fault trees (DFT) are introduced as an extension of FTA and as an example of domain-specific analysis. Section 3 shows a use case from the field of machine safety. Finally, Sect. 4 draws a conclusion and gives an overview of future developments.

2   General Approach

2.1   Goals

The general method presented in this paper develops modeling extensions for the practical applicability of the RAAML approach in domain-specific structures, using established methods and SysML as a common platform. On the modeling side, profile extensions continue to provide tight model coupling. The execution of analyses takes place in external analysis tools to make the approach more accessible to small and medium-sized enterprises (SMEs) with their limited resources, established structures and proven-in-use tools. Translators are developed to further leverage existing analysis methods and tools to convert the with profiles enriched models into the model-specific representation of a certified analysis tool. For feedback, the translator can also merge the results back into the SysML models to identify unmet requirements.

2.2   Modeling Extensions

RAAML describes the basis for modeling and explaining the methodology. It was chosen because of its easy extensibility to other domains via general base profiles. It enables the simple extension of static influencing factors for dynamic models and domain-specific analyses. However, the approach presented here is based on loose tool coupling. In this way, existing mature tools of certain domains and their licenses can continue to be used. In addition, complex algorithms and calculations do not have to be integrated and validated in the profiles at great expense. Analysis results can be automatically fed back into system models, promoting consistency. This approach is a generally valid adaptation and extension of the OMG approach. The profiles are based here on the domain of machine safety and its domain-specific analysis but can be designed similarly for other domains as well.

The extensions present time-dependent basic events and the concept of integrating dynamic fault trees (DFTs) as profile extensions of the FTA method. DFT represents a different analysis method based on FTA but dramatically extends its functionality through dynamic gates. Their origin can be found in Dugan et al. [5], in which a large number of variants of DFTs were proposed. An overview of the extensions can be found in [7]. The structure of DFTs is extended compared to static fault trees; additional gates are introduced to overcome the latter limitations. The new gate types are Priority-AND (PAND), Priority-OR gate (POR), Sequence-Enforcing (SEQ), Functional Dependency Gate (FDEP), Cold Standby Spare (CSP), Warm Standby Spare (WSP), and Hot Standby Spare (HSP). PAND and POR are special dynamic versions of AND and OR with a prioritization. An input event A must occur before an input event B. SEQ defines a sequential fault behavior with its priority values and FDEP defines a functional dependency of different events on a trigger event. CSP, WSP and HSP consist of a primary component and several spare components that can be activated in the event of a failure of the primary component. A detailed functional description can be found in [5]. DFTs also include time-continuous calculations, which consider the evolution of system failures over time. The failure behavior of components is usually specified by a probability function that gives, for each point in time, the probability that the component has not failed yet [7]. Additionally, if components can be repaired without affecting the operation of other components, the events have an additional repair time distribution. Another essential factor, especially for SPARE gates, is a dormant phase when a component is not in use or is redundant [5]. Failure distributions can often be adequately approximated by inverse exponential distributions, as described in IEC 61508 or [7]. However, other distributions exist depending on the domain, such as Weibull or LogNormal Distribution. Continuous-Time-Markov-Chains (CTMC) can offer a detailed and precise calculation possibility for DFTs [5]. DFTCalc [Arnold et al. 'DFTCalc: a tool for efficient fault tree analysis' in 2013] and Stormchecker [Hensel et al. 'The probabilistic model checker Storm' in 2021] have been identified as suitable open-source tool environments for DFTs. Both support efficient modeling through compact representations, practical analysis over wide ranges of reliability properties, efficient analysis via stochastic techniques, and CTMC.

Based on [6], within the existing package "Methods", this paper extends the package "FTA" and creates a package "DFT". "DFT" defines the extensions based on the "FTA library". The "DFT Profile" defines the dynamic gates, events, and trees' stereotypes related to the "FTA Profile". The "DFT Library" uses these stereotypes to specify the behavior and parameters, as depicted in Fig. 1. The extension concepts for basic events were integrated into the profile extensions based on IEC 61508-6 and IEC 61025 [1]. In the left part of Fig. 1, the extension redefines the constant "Probability" by the "failureRate", "failure-Lambda" or "failureMean" depending on the distribution function. Additionally, "dormancy" and "coverage" are introduced for every dynamic basic event. Every "DFTree" has a "failureExposureTime" defined for the whole fault tree. For a "Repairable - EBE" the "repairRate" and for "Latent - EBE" the "inspectionInterval" must be specified additionally.

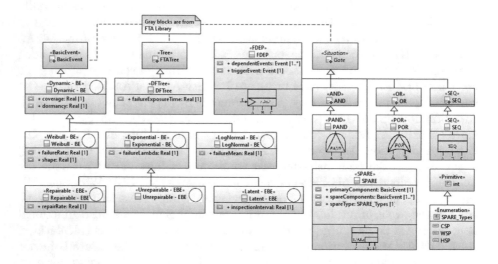

Fig. 1. DFT Library Extensions for Events and Gates

Temporal and sequential dependencies are also found in gate behavior in the right part of Fig. 1. The PAND, POR and SEQ gates have been described with the "priority" property prepared by the OMG in the FTA library events. Thus, no change is made except for the redefinition. For the SPARE gate, the approach redefines the "sourceGates" for the "primaryComponent" and "spare-Components" properties. The property "spareType" defines the configuration as CSP, WSP and HSP. A FDEP gate has one "triggerEvent" input and several "dependentEvents" outputs.

2.3   Analysis Tool Integration

SysML modeling documents based on XML Metadata Interchange (XMI) standard provide the basis for easy export to external analysis tools and feedback of the results. Using the prevailing know-how, developers can develop transformers and connect already established, validated and certified analysis tools. Eclipse Modeling Framework (EMF) offers a very integrative approach to transformation development in conjunction with UML and XML. However, the knowledge of National Instruments LabVIEW (LabVIEW) plays a central role in the machine industry and the test environments established there. It is often used as a rapid prototyping tool in the machinery domain. LabVIEW offers library functions that can read and interpret XML files.

In the same way as the OMG approach, simple navigation between the system model and the analysis model, simple maintenance of the analysis model in case of system model changes, and the ability to build the analysis model at the system level arise in this approach [6]. However, unlike the OMG approach, performing the analyses not directly in the models contributes to a looser linkage between the system model and the analysis model. RAAML allows entities to be better traced, such as the relationships between and effects on calculations. If there are references between entities from the analysis algorithms, they cannot be considered in this approach. In contrast to the OMG approach, this offers a concept that is easier to extend for SMEs.

2.4 Implementation Status

The first step was the integration of the OMG RAAML profiles into a modeling environment, here Eclipse Papyrus. The concepts for integrating time-dependent basic events, dynamic gates, and associated DFTs were then developed. Next, translators were implemented in LabVIEW to convert the enriched models via XMI into the representation of different analysis tool, namely RAM Commander from ALD, and Stormchecker, as well as to re-import the analysis results. A translator for the ALD Fault Tree Analyzer in RAM Commander was first written to prove the concept. It consists of 4400 lines of LabVIEW code. Due to the minimal cutset of calculations in the tool, however, substantial calculation deviations arose for DFTs, so the RAM Commander has proven impractical for this context. Finally, the Stormchecker tool was chosen for the final translator. Due to the knowledge gained from the previous translator and the easy to use Galileo input format of Stormchecker, the new translator could be implemented with only about 2200 lines of code. The development time of fewer than three weeks demonstrates the rapid prototyping capabilities of NI LabVIEW.

2.5 Related Work

Researchers developed various approaches that use MBSE to integrate and automate artifacts for safety analysis methods. Yakymets et al. [8] built the system architecture using SysML and converted the SysML model into an AltaRica model, which generates the fault tree for each hazard. These make synchronization between stakeholders difficult because of different modeling languages. To circumvent the limitations of loose model coupling and avoid data loss, researchers developed the idea of tight model coupling to analysis models. This allows stakeholders to integrate safety-relevant artifacts into SysML via profiles. The OMG approach with tight model coupling and tool coupling provides the basis for modeling a variety of safety and reliability artifacts that are tightly coupled to the system design [5]. In contrast to the OMG approach, many application-based approaches have tended to be limited to loose tool coupling. However, there is no approach applicable and extensible for a variety of safety and reliability analyses like RAAML. For example, Helle [6] developed a method to generate Reliability Block Diagrams (RBD) connected to a Java program, henceforth called SafetyAnalyzer. He integrated safety artifacts, such as failure rates of components, using so-called tagged values. Even though the method is limited to reliability analysis, it showed that a safety analysis could be performed automatically from SysML models. The authors of [4] and [4] have already addressed the integration of hazard analysis using SysML profiles. Mhenni et al. [4] used SysML profiles to capture safety-related information for a model-based safety analysis, which can generate safety artifacts at each stage of the safety lifecycle. The approach already considers FMEA and FTA, but is limited to them. Baklouti et al. [9] is an extension of [4] and additionally captures redundancies and generates DFTs with loose tool coupling. Biggs et al. [4] then created a SysML profile called SafeML for reliability analysis. It already describes a similar approach to RAAML, but lacks abstracting profiles, such as Core and General, and uses loose tool coupling. The approach presented in this paper therefore creates a methodology to extend the universal approach of RAAML for the implementation of a concrete domain-specific extension. Based on the last mentioned publications it reuses established domain-specific tool environments.

3   Use Case

The approach is demonstrated using a small example regarding machine safety.

3.1   Application Model and Safety Requirements

The example consists of an emergency venting system, depicted in Fig. 2. When detecting escaping gas, a suction system must be activated that extracts the toxic gases. A main suction system "main_suction" is supported by a redundantly installed backup suction system.
"backup_suction". A main gas detection sensor "main_sensor" and a comparison sensor "compare_sensor" detect the escaping gas. "compare_sensor" maintains the function if the "main_sensor" fails. If both fail, an operator can press the emergency stop button "emergencyStop" and activate the suction systems.

The values measured by the sensors are compared and checked for plausibility by the main safety controller "main_cu". From experience, it is assumed that a single controller is insufficient for the desired failure probability, which is why a second safety controller "monitoring_backup_cu" is installed. All components are assumed to be non-repairable with constant failure rates from $5,5*10^{-7}$ to $5,5*10^{-5}$ and exponential distribution depicted in the basic events of Fig. 3. The system is required to be operable for at least 36,000 h (4 years 24/7) with a requested Performance Level PLr= e (SIL3) according to ISO 13849-1, i.e., with an probability of failure (PF) less than $3.6*10^{-3}$. An analysis must show whether

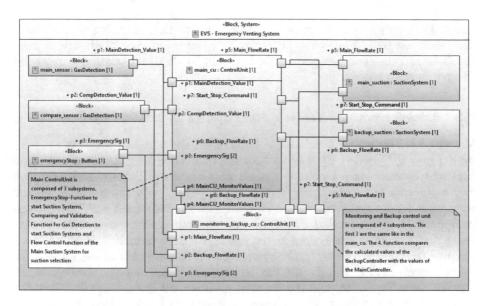

Fig. 2. Use Case System Model

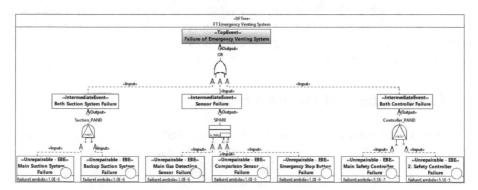

Fig. 3. Use Case Dynamic Fault Tree

the system can meet the specified PF with the modeled components and their failure rates (FR).

Based on the system model, the corresponding dynamic fault tree (DFT) can be found in Fig. 3. The requirements describe that the complete system fails in case of the failure of both controllers, sensors, or suction systems. Between the "Main Suction System Failure" and the "2. Suction System Failure", as well as between the "Main Safety Controller Failure" and the "2. Safety Controller Failure" there are thus PANDs modeled. In addition, there is a hot spare between the sensors because the sensors partly run in parallel. However, without the failure of the Main Gas Detection Sensor, the other components do not play a supporting role and can be replaced at any time. In order to create a reference to the system model the safety artifacts can be connected to each element using dependencies like "RelevantTo", "Violates", "Detection", "Prevention" and "Mitigation" from the RAAML core and general profile. Its main use is to illustrate dependencies to system components.

3.2 Transformation, Analysis and Results at Model Level

Via the XMI transformer written in LabVIEW, the enriched artifacts are converted into the Galileo input format for the Stormchecker evaluation tool. The tool was invoked via a Docker container and command line in LabVIEW and had a runtime of 0.344 s. After back transformation onto the model level, the computed failure probabilities are stored in attributes of the DFT nodes and compared to the requested PF. With the current input, the analysis results show that the PF of the system at the selected failure rates is $6.73*10^-3$ after a runtime of 36000 h. Therefore, the system would only be able to run for 28000 h for the requested PF. In this case the requirement is not met and the coloring of the DFT, its TopEvent, and the violated requirement change to red. Since the failure of one controller has the highest probability of a dangerous failure, it is decided by the stakeholders to use a different backup controller with a higher FR of $5.5*10^-6$ to reduce the PF of the overall system. A new calculation results in a PF of $2.5885*10^-3$ after 36000 h. With the re-import of the results, the TopEvent that is no longer violated changes to green, indicating that the requirement of the TopEvent is now met.

4   Conclusion and Future Work

The presented approach describes a practice-oriented procedure and implementation for using safety and reliability analyses in MBSE. Furthermore, based on RAAML, an extension and adaptation for DFTs have been implemented, which, like RAAML, is based on tight model coupling, but unlike RAAML, is based on loose tool coupling. The DFT profile and library allow modeling of time-dependent probabilities, dynamic correlations, and their propagation in the fault tree. The profiles created can be used for machine safety and other domains. Unlike RAAML, the loose coupling at the tool level mean that relationships to parameters cannot be tracked during calculations and computations within parameter diagrams. However, in areas such as the machine industry, established tools are essential. Linking MBSE and safety analysis in SysML and reusing established tools practically bridge the gap between system design and safety analysis for SMEs in machine industry. The market's need for custom products and special-purpose engineering allows SMEs with flexible structures to gain better access to the market for safety-critical applications.

Further work needs to validate the presented approach for more complex systems. The profiles defined by RAAML and already extended and adapted in this paper can be extended for further analyses based only on "Core" and "General". Analyses commonly used in the machine industry can be found in the risk graph method, which works similarly to FMEA, or the block diagram method from ISO 13849 using the SISTEMA analysis tool. A process for evaluating the approach with an SME as a partner is currently underway.

Acknowledgments. Many thanks to Prof. Dr. Reinhold Kröger for his invaluable advice and support.

References

1. Ruijters, E. and Stoelinga, M.: Fault tree analysis: A survey of the state-of-the-art in modeling, analysis and tools, Computer Science Review, 2015
2. OMG: Information technology - OMG SysML, 2017
3. Berres, A., Post, K., Armonas, A., Hecht, M., Juknevičius, T. and Banham. D.: OMG RAAML standard for model-based Fault Tree Analysis, INCOSE International Symposium, (2021) 1349–1362
4. Mhenni, F., Nguyen, N. and Choley, J.-Y.: Automatic fault tree generation from SysML system models, IEEE/ASME, (2014) 715–720
5. Yakymets, N., Sango, M., Dhouib, S., Gelin, R.: Model-Based Engineering, pp. 6136–6141. IEEE/RSJ IROS, Safety Analysis and Risk Assessment for Personal Care Robots (2018)
6. Helle, P.: Automatic SysML-based safety analysis, ACES-MB, 2012
7. IEC 61508:2010: Functional safety of electrical/electronic/programmable electronic safety-related systems, 2010
8. Junges, S., Guck, D., Katoen, J.-P. and Stoelinga, M.: Uncovering Dynamic Fault Trees, 46th Annual IEEE/IFIP DSN, (2016) 299–310
9. Mhenni, F., Nguyen, N. and Choley, J.-Y.: Automatic fault tree generation from SysML system models, IEEE/ASME, (2014) 715–720
10. OMG: Information technology - OMG SysML, 2017
11. Ruijters, E. and Stoelinga, M.: Fault tree analysis: A survey of the state-of-the-art in modeling, analysis and tools, Computer Science Review, 2015
12. Yakymets, N., Sango, M., Dhouib, S., Gelin, R.: Model-Based Engineering, pp. 6136–6141. IEEE/RSJ IROS, Safety Analysis and Risk Assessment for Personal Care Robots (2018)

Software-Defined Networking with Prioritization for a Redundant Network Topology Called Double Wheel Topology

Dimitrios Savvidis[(✉)], Janis Marrek, and Dietmar Tutsch

University of Wuppertal, Chair of Automation / Computer Science,
Rainer-Gruenter-Str. 2, 42119 Wuppertal, Germany
savvidis@uni-wuppertal.de

Abstract. In this paper we describe SDN and apply it to our double wheel topology. With the double wheel topology, new network features are available that can be used for automation in everyday life. By using SDN we show the reduction of latency between two hosts by OpenFlow on the double wheel topology.

Keywords: SDN · OpenFlow · Ryu · topology · wheel · latency · real-time

1 Introduction

In daily life, more and more devices have a network interface and can be used for automation in everyday life. For this automation, the systems must be able to operate in soft real-time and communicate with each other. More and more devices for the automated home are being given a form of "intelligence" for this purpose and are following the trend for cloud applications. This intelligence is controlled by the software, which also determines the life cycle of the hardware and can also make it useless. The software defines the functionalities and integration in the automation in everyday life. This concept is called Software-Defined System (SDS). One of the characteristics of SDS is Software-Defined Networking (SDN), which is a new networking concept and enables new networking scenarios. SDN decouples the network hardware from its software. The network is controlled by an SDN controller in the *Control Plane* (CP). The CP takes over the software functionality of the network hardware. The network hardware is responsible for transmission and is located in the *Data Plane* (DP) and is controlled by the CP [1]. The decoupling of hardware and software enables new dynamic implementations that are useful for communication in everyday automated applications. In this paper, we introduce a new topology and their potential by using SDN. For this purpose, we use a network topology consisting of a star and two rings, a hybrid topology we have called *double wheel topology*

H. Unger and M. Schaible (Eds.): Real-Time 2022, LNNS 674, pp. 82–87, 2023.
https://doi.org/10.1007/978-3-031-32700-1_9

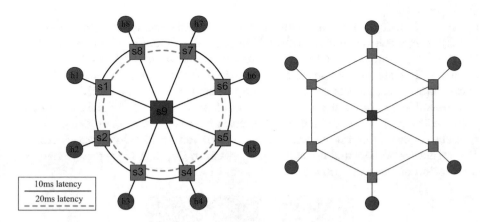

Fig. 1. double wheel topology   **Fig. 2.** wheel topology

which can be seen in Fig. 1. The double wheel topology is a modification of the *wheel topology* (see Fig. 2 and [2]). By using SDN with this hybrid topology, we want to reduce the latency in this redundant network topology and also show that topologies that normally would be disadvantageous for network use can be used in a new way with SDN in an automated home. We show in a simulation how SDN and the double wheel topology can be used to optimize network traffic that can be used for automation in everyday life. We measure the latency between two hosts with and without "latency prioritization" and give an outlook on the use of wheel topologies with SDN in different use cases.

2 Basic SDN Requirements

To implement an SDN based network architecture the following components are needed: an SDN Controller, an SDN compatible switch and a corresponding southbound API protocol to configure the SDN switches. The SDN controller is located in the control plane and controls the network. There are various SDN controllers, each with different functional characteristics. In this work the Ryu SDN Controller serves as SDN Controller. The Ryu SDN Controller is a controller based on the python programming language. The code of Ryu is open source under the Apache 2.0 license. Ryu can use different protocols, especially Open-Flow. OpenFlow is used as the southbound API protocol to generate behavior rules for the network packets. These behavior rules are called flows and are stored in flow tables on the SDN switches. When a network packet arrives at an SDN switch, the SDN switch checks the flow tables for a matching entry and processes the packet. If no matching entry is available, the SDN switch requests a procedure from the SDN controller. Because the SDN network architecture in this work is only created virtually, Mininet is used as network emulator. Mininet is a network emulator that allows to create virtual hosts, switches and connections under Linux. Using Mininet it is possible to simulate an SDN compatible network infrastructure, like switches and hosts in the Data Plane and connect them to an SDN Controller in the Control Plane. Mininet can create Open vSwitch (OVS) compatible virtual switches which are fully OpenFlow compatible. The Control Plane contains two main interfaces, the southbound API and the northbound API. The southbound API is responsible for the communication with the Data Plane layer, i.e. the communication to the switches. The northbound API is responsible for the connection and communication with the Application Plane, with various software based network applications that would otherwise be found in a classic switch or router, i.e. routing algorithms or MAC-Address learning. The structure of SDN and its layers can be seen in Fig. 3. A more comprehensive description of the SDN concept can be found in [1].

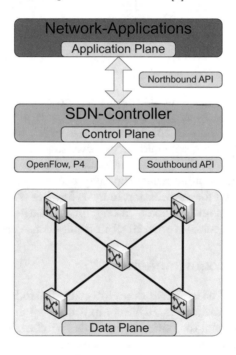

Fig. 3. SDN layer

3   Realization

To demonstrate the double wheel topology, we build an SDN network using OpenFlow as Southbound API to apply flows for latency prioritization between hosts. We use the Ryu SDN Controller and Mininet to create virtual OpenFlow compatible switches and simulate the double wheel topology in a network. The double wheel topology consists of one star and two rings and is a modification of the wheel topology consisting of one star and one ring. Both topologies differ from the other topologies by the fact that the nodes are connected to the switches on the ring. From now on the nodes will be called hosts. We use the double wheel topology with nine switches (s1 to s9) and eight hosts (h1 to h8) (see Fig. 1).

Each host (h1 to h8) is connected to one of the eight switches (s1 to s8). With these switches (s1 to s8) we form two parallel rings. We use a "fast" ring, where the hosts and the switches have a connection with a 10 ms latency, and a "slow" ring, where the switches have another parallel connection with a 20 ms latency. For the embedded star network, an additional "central" switch (s9) is required. The central switch has a connection to every other switch with a latency of 10 ms.

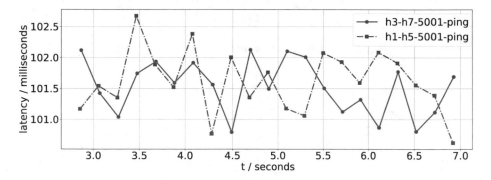

Fig. 4. sockperf measure without latency prioritization

The current network contains a redundant connection due to the additional star network, which could lead to loops and potentially cause the network to collide. By using Multi-Path-Routing (MPR) with the Ryu controller, collision free paths are created without disabling connections between end devices as in the classic network by the Spanning-Tree-Protocol (STP). With MPR activated, several paths can be used at the same time which primarily leads to a bandwidth increase. This is shown in [3]. After the implementation of the network the Ryu controller has to be trained by means of the network. By performing pings between each host, Ryu learns and computes the necessary paths as flows in the OVS switches s1 to s9. After all flows are computed, a latency measurement is performed using the *sockperf* tool. For this purpose we establish two parallel latency measurements between the hosts h3 and h7 as well as h1 and h5. Because MPR is activated by default, a path over the "slow" ring is selected and a latency measurement of about 100ms is calculated, see Fig. 4. Because of the simple design of our topology we can verify the previously measured latency by simply counting the hops, see Fig. 1. Furthermore we create a latency prioritized flow for a UDP connection between h1 and h5 on port 5002 and pass it to the Ryu controller, which installs the necessary flows on the switches. With this new "flow latency prioritization" a new path between h1 an h5 is used. Now, a new *sockperf* measurement utilizes the lowest latency path over the "fast" ring. The *sockperf* measurement provides a latency of about 40ms, see Fig. 5. We can now successfully verify these measurements again by simply counting the hops on the shortest and direct path between h1 and h5, see Fig. 1.

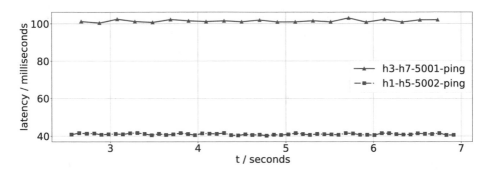

Fig. 5. sockperf measure with latency prioritization

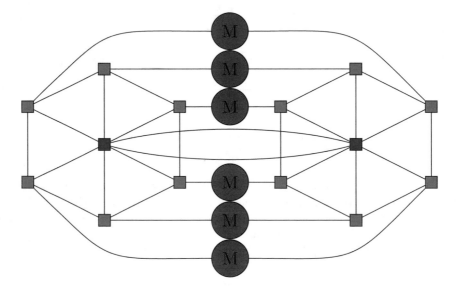

Fig. 6. mirror wheel topology

4   Conclusion and Outlook

In this scenario, we have shown the use of the double wheel topology with SDN. Using the possibilities that SDN provides, we have implemented latency prioritized flows with the Ryu SDN controller that has allowed us to reduce the latency between hosts through an optimized path. The use of SDN opens new perspectives and possibilities for network topologies that are not used in classic networks for their known disadvantages. By using the wheel topology and double wheel topology, we have a topology that makes it possible to achieve a latency reduction through SDN. The wheel topology can be used in special dedicated networks such as automation in everyday life for the realization of soft real-time. By using prioritized flows, soft real-time can be achieved for individual hosts. For further research we will investigate the wheel topology in further perspective, using different tools to analyze their network behavior in details. We will also investigate their extension for a redundant host connection, such as the mirror wheel topology (see Fig. 6). Furthermore, we would like to check more SDN controllers and consider the combination of multiple SDN controllers. In addition, we would like use the P4 network programming language to test different use cases of SDN with a wheel-based topology. Another possible application scenario of the wheel topology is a modified version for an automotive network, which is also a dedicated network. For this purpose an investigation based on SDN using the Time-Sensitive Networking (TSN) standard can be performed to fulfill the critical requirements in real-time, as it is shown in the work of Häckel et al. in [4] and of Çakir et al. in [5].

References

1. Kreutz, D., Ramos, F.M.V., Veríssimo, P.E., Rothenberg, C.E., Azodolmolky, S., Uhlig, S.: Software-defined networking: a comprehensive survey. Proc. IEEE **103**(1), 14–76 (2015). https://doi.org/10.1109/JPROC.2014.2371999
2. Savvidis, D., Roth, R., Tutsch, D.: Static evaluation of a wheel-topology for an SDN-based network Usecase. In: Würzburg Workshop on Next-Generation Communication Networks (WueWoWas 2022) (2022). https://doi.org/10.25972/OPUS-28071, urn:nbn:de:bvb:20-opus-280715
3. Yahya, W., Basuki, A., Maulana, W., Akbar, S.R., Bhawiyuga, A.: Improving end-to-end network throughput using multiple best paths routing in software defined networking. In: 2018 10th International Conference on Information Technology and Electrical Engineering (ICITEE), pp. 187-191 (2018). https://doi.org/10.1109/ICITEED.2018.8534837
4. Hackel, T., Meyer, P., Korf, F., Schmidt, T.C.: Software-defined networks supporting time-sensitive in-vehicular communication. In: 2019 IEEE 89th Vehicular Technology Conference (VTC2019-Spring), pp. 1–5 (2019). https://doi.org/10.1109/VTCSpring.2019.8746473
5. Çakir, M., Häckel, T., Reider, S., Meyer, P., Korf, F., Schmidt, T.C.: A QoS aware approach to service-oriented communication in future automotive networks. IEEE Veh. Netw. Conf. (VNC), 1–8 (2019). https://doi.org/10.1109/VNC48660.2019.9062794

A Survey on Pedestrian Detection: Towards Integrating Vulnerable Road Users into Sensor Networks

Maximilian De Muirier[(✉)], Stephan Pareigis, and Tim Tiedemann

Department of Computer Science, Hamburg University of Applied Sciences,
Berliner Tor 7, 20099 Hamburg, Germany
`maximilian.demuirier@haw-hamburg.de`

Abstract. This paper provides an overview of fundamental and recent 2D and 3D pedestrian detection methods. It is part of the ongoing investigation on integrating vulnerable road users into sensor networks. Besides the sensor-specific object detection methods based on LIDAR sensors, RADAR sensors, thermal imaging cameras, RGBD cameras, and RGB cameras, a selection of sensor fusion methods is presented. Shown methods have been developed to increase traffic safety for vulnerable road users, which include pedestrians and cyclists.

Keywords: computer vision · pedestrian detection · sensor fusion · 2D · 3D · lidar · radar · rgb · thermal · vulnerable road users

1 Introduction

The increase in inner-city traffic due to urbanization and demographic change is leading to a rise in the use of alternative means of transport, such as bicycles and e-scooters. Unclear and unpredictable traffic is an adverse effect that results in more and more accidents among various road users, to the dismay of the most vulnerable ones [1]. Vulnerable road users [2], which include pedestrians, cyclists, elderly, children and many other road users, can be reliably localized and identified with spatial precision using multimodal sensor networks combined with state-of-the-art ML-based 3D pedestrian detection. The joint use of LiDAR and RADAR sensors, as well as RGB and thermal imaging cameras, makes pedestrian detection less susceptible to weather and lighting conditions [3]. Performance and precision of applied methods can be evaluated using test data sets such as the popular KITTI benchmark [4].

Sensor fusions can compensate sensor-specific disadvantages and combine advantages but with the cost of methodological complexity. A data-driven and multimodal pedestrian detection approach can reduce such described traffic accidents in future.

Section 2 deals with the 2D and 3D pedestrian detection methods based on the sensor technology.

H. Unger and M. Schaible (Eds.): Real-Time 2022, LNNS 674, pp. 88–96, 2023.
https://doi.org/10.1007/978-3-031-32700-1_10

Section 3 addresses the topic of 2D and 3D sensor fusion methods and describes typical sensor combinations.

Section 4 gives a conclusion and outlook.

2   Methods for 2D and 3D Pedestrian Detection

This chapter deals with the different sensor types, their advantages and disadvantages, and applied methods for pedestrian detection in more detail. The following pedestrian detection methods are also used to detect cyclists. The object detection methods referenced in the scope of this paper are considered in terms of the detection of pedestrians and cyclists.

2.1   RGB Camera

2D pedestrian detection started with RGB cameras. These are available in a range of resolutions, frame rates and at different price points. Their use is strongly dependent on weather and light conditions. Therefore, rain, fog or darkness make their use practically futile. RGB cameras produce 2D pixel-based images. ML-based 2D pedestrian detection is realized using different methods such as YOLO [5,6], SSD [7] or Faster R-CNN [8]. The 2D object detection must be done in two stages. The first stage involves the generation of a 2D bounding box of a detected object within the RGB image. This procedure can also be done separately in each image channel and then combined afterwards. In the second stage, a class is assigned to the detected object, such as a pedestrian or cyclist, with a confidence probability. These two stages do not necessarily have to be done sequentially. Similarly, the pixels associated with the detected object can be segmented [9]. Spatial estimation of the detected object solely with RGB data is not possible.

Two-Stage Methods

Two-stage object detection methods are commonly used as 2D object detectors and classifiers. In the first stage, a region proposal network (RPN) suggests 2D candidate proposals, which are then handed over to the classifier within the second stage after being cropped and resized. The advantage of such a procedure is a significantly higher detection precision. The initial R-CNN inherited the slow inference and computational intensity of classical CNNs and was superseded by Fast R-CNN [10] and Faster R-CNN. The first made end-to-end training possible and added bounding box regression. Faster R-CNN added a small CNN called Regional Proposal Network (RPN) that can be found in many other recent 3D pedestrian detection methods.

One-Stage Methods

Liu et al. [7] presents SSD. In such a single-stage architecture, a set of dense anchor boxes is regressed and classified in a single step, resulting in a class prediction per anchor box. This detection method is simple and fast. Single-stage caught up with other more complex two-stage object detection methods by further implementing a focal loss function [11] for better prediction.

Redmon et al. [5] designed YOLO as a single-stage end-to-end object detection method that generates Region-of-Interests (RoI) within 2D data similar to R-CNN. However, YOLO differs from other methods in that input data is processed only once. This fact makes YOLO very fast and thus suitable for real-time applications. Of course, this limitation also has disadvantages, such as comparatively worse class prediction. Compared to Faster R-CNN, YOLO makes fewer background errors but more localization errors. Newer versions have led to significant improvements in these areas [6].

He et al. [9] introduces Mask R-CNN as a state-of-the-art method for segmentation. Besides 2D object localization and classification, this method distinguishes which pixels actually belong to the corresponding detected class object. This binary mask is outputted for each Region-of-Interest (RoI) parallel to the prediction of the class and anchor box. The segmentation process is enabled by learning object masks.

2.2   Thermal Imaging Camera

Thermal imaging cameras produce a pixel-based 2D mono image in which each pixel is assigned to a corresponding normalized temperature as a value. Pedestrian detection using thermography is largely independent of weather and lighting conditions. Consequently, its use has a clear advantage over conventional RGB cameras. However, noisy 2D mono images suffer in terms of contrast and texture, comparatively. Since thermal imaging cameras measure the radiation of pedestrians in the infrared spectrum, the success of the detection depends on the temperature difference to the immediate environment.

For 2D pedestrian detection based on infrared images, the same methods as those based on RGB images can be applied. These methods are based on CNN frameworks such as YOLO [5] or SSD [7]. The pedestrian detection steps, therefore, consist of feature extraction, bounding box regression and object classification.

You et al. [3] propose TIPYOLO, a real-time pedestrian detection method optimized for edge computing devices. A multi-scale mosaic data augmentation is used to highlight the diversity and texture of objects. Reducing the number of features the framework can extract minimizes the memory requirements and the computational costs.

2.3   RGBD Camera

An RGBD camera is a type of depth camera that provides both RGB data and corresponding depth data. The data are synchronized and aligned with each other. The depth map is generated using the stereoscopic effect when two rectified cameras are placed at a fixed distance from each other. The resolution of the depth image is based on that of the RGB image. It has the same operational limitations as an RGB camera. The colored point cloud resulting from the fusion of RGB image and depth map can be used for both 2D and 3D object detection.

Qi et al. [12] fed both 3D point cloud from LiDAR sensors and RGBD data into F-PointNet. Further information can be found in Chap. 3.1. For this, the depth map ist first converted to a point cloud. Then regional proposals are generated in parallel from the 2D image and 3D point cloud. Each 2D regional proposal is then extruded to a 3D viewing frustum, which is then used for 3D object detection within the spatially reduced point cloud.

2.4 LiDAR Sensor

The previous chapter has shown, depth perception offers a distinct advantage in pedestrian detection that 2D sensors such as an RGB or thermal imaging camera cannot provide. Since the depth camera's detection range is limited due to its pixel-based functionality, there is a need for sensors with a more extended range. The LiDAR sensor has a several times greater range than said RGBD cameras and can detect pedestrians at much longer distances. They are also not affected by weather or lighting conditions, similar to thermal imaging cameras. These sensors differ in field of view, detection distance, horizontal and vertical resolution, and frame rate. Sensors with good ratings in all of these categories tend to be expensive. Due to their vertical field of view and associated number of horizontal scan layers, objects in the distance are displayed with lower vertical resolution. This characteristic negatively affects the quality of pedestrian detection. If the LiDAR sensor is on a moving object, the adverse effects of the rolling shutter are added. Solid-state LiDAR sensors do not have this disadvantage due to their global shutter, which makes them preferable. LiDAR sensors store their data in sparse and partially unstructured point clouds. The spatial representation of these point clouds is usually based on points, voxels or multi-view.

Point-Based Methods

Qi et al. [13] present PointNet as the first point-based object detection method that can directly extract spatial features from raw point clouds based on each point, forming a global feature group. These global features are then used for classification and segmentation. Unlike voxelization [14], see Chap. 2.4, and 2D projection [15], see Chap. 2.4, it does not involve any kind of loss. PointNet scales with the size of the point cloud losslessly. Qi et al. subsequently developed PointNet++ [16], that uses besides the global parallel also local information by means of a multi-scale and multi-resolution grouping. Moreover, improvements were done in parallelizability and data structure.

Shi et al. [17] introduce PointPCNN, a two-step 3D object detection method based on PointNet++. 3D regional proposals are generated using a bottom-up method in the first stage. A mask segmentation generates foreground points that are processed first. In the second stage, these bounding box proposals are refined and point representations are learned from the first stage using regional pooling.

Voxel-Based Methods

Zhou et al. [14] introduces the VoxelNet as the first object detection method that divides the 3D space into voxels to extract local features directly from the raw point cloud end-to-end. Voxels are grid-based volumetric representations of the 3D space and can thus group features of individual points. Voxelization comes with a loss of precision with respect to the particular locations of the points. Each voxel is treated as a separate PointNet.

Similarly, Lang et al. [18] presents PointPillars that groups the voxelized point cloud into columns called pillars.

With Voxel R-CNN, Deng et al. [19] combined the advantages of a Regional Proposal Network (RPN) with subsequent voxelized Region-of-Interests (RoI) with those of 3D feature extraction. To do this, it projects the voxels into the bird's eye view (BEV), from which the 3D regional proposals are then generated.

Point-Voxel Fusion-Based Methods

The combination of voxel-based and point-based 3D object detection methods allows for jointly extracting features and thus combining the advantages of both general approaches.

Li et al. [20] propose P2V-RCNN, a point-to-voxel feature learning method that voxelizes the raw 3D point cloud but extracts point-wise semantic and local spatial features. A voxel-based Regional Proposal Network (RPN) then provides 3D object proposals.

Shi et al. [21] additionally introduce voxel-to-keypoint scenes with PV-RCNN. Here, multi-scale features of a whole scene are abstracted to a small set of keypoint features. These keypoint features maintain exact locations. In their follow-up method [22], more keypoint features are extracted and resource consumption is reduced through an improved pooling process.

Wang et al. [23] implement a weighting component for the semantic and geometric part of a fused extracted feature in their 3D object detection method PVFusion.

Multi-view-Based Methods

Image-based or multi-view-based 3D pedestrian detection methods generally use 3D data but first project them into the 2D view. The bird's eye (BEV) or front view is often used for the generation of candidate proposals in a first step.

Chen et al. [15] created MV3D that initially projects the 3D point cloud into the bird's eye view (BEV) to generate accurate 3D object proposals. The advantage is that these 3D proposals can be projected effortlessly into any other 2D view. Faster-R-CNN is then used to learn region-wise features from multiple 2D perspectives of the point cloud. This type of technique is referred to as multi-view.

2.5   RADAR Sensor

RADAR sensors are now an integral part of the sensor technology of a modern automotive vehicle with a degree of autonomy. In addition to the RADAR cross-section and the angular distance, they also provide a Doppler velocity, which describes the relative velocity of the detected object. Compared to LiDAR sensors, RADAR systems have a higher frame rate, but a lower horizontal and vertical resolution. RADAR sensors provide a point cloud that describes the distance of detected objects in Cartesian coordinates.

Cennamo et al. [24] developed a method using PointNet++ [16] to detect pedestrians in 3D space using only RADAR sensors. This method relies on measuring the Doppler velocity of detected objects and distinguishing them from the static environment.

3   Sensor Fusion Methods for 2D and 3D Pedestrian Detection

In the previous chapters, methods for pedestrian detection are presented, each based on only one type of sensor. It was demonstrated that these methods are able to reliably and accurately localize and classify pedestrians and cyclists in either 2D or 3D space. The limitations of the used sensor are also that of the detection. A sensor fusion balances out the advantages and disadvantages of each sensor. In the following chapter, four different sensor combinations are presented, and the underlying methods for pedestrian detection are explained.

3.1   LiDAR Sensor and RGB Camera

Frustum-Based Methods
Frustum-based pedestrian detection methods such as F-PointNet from Qi et al. [12] and F-ConvNet from Wang et al. [25] take advantage of the already widely known and well-proven 2D pedestrian detection based on YOLO [6], SSD [7] and Faster R-CNN [8]. In the first step, these methods use RGB data to generate 2D regional proposals for pedestrians, which again are projected into the 3D space of the point cloud. This step serves as a pre-filter for the more targeted processing of the 3D point cloud. These Regions-of-Interest (RoIs) are subsequently the starting point for 3D pedestrian detection in the second step. Reducing the number of points to be processed significantly minimizes the search space, the memory footprint, and the inference time. These properties make such methods suitable for real-time applications. However, the success of 3D pedestrian detection strongly depends on that of the preceding 2D detection methods.

Multi-view-Based Methods
In contrast to MV3D, which is based only on LiDAR data, other methods additionally use RGB data.

Ku et al. [26] propose with AVOD a method that also projects the point cloud into the bird's eye (BEV) and front view for feature extraction. Likewise, features are extracted from the RGB data and then fused with those of the front view by a 3D Regional Proposal Network (RPN). This approach enables an even more accurate 3D pedestrian localization.

3.2   RADAR Sensor and RGB Camera

Similar to Chap. 3.1, 2D and 3D data are merged. The methods for 2D and 3D pedestrian detection work in the same way. However, temporal information provided by the RADAR point cloud can be used to separate detected objects from the static environment.

Dimitrievski et al. [27] exploit this advantage in their implementation of a spatio-temporal CNN. First, this neural network is trained to distinguish objects and ground using a weakly supervised deep learning model. The radar network is then trained to simultaneously estimate both the position and the class of the detected objects. The RGB data is used by a Regional Proposal Network (RPN) to generate 2D regional proposals.

3.3   LiDAR and RADAR Sensors

The fusion of two 3D sensors, such as LiDAR and RADAR sensors, combines their respective strengths in 3D pedestrian detection. Both sensors generate point clouds that can be used to generate 3D bounding box proposals. Also, a certain redundancy in the safety aspect is of advantage. In contrast to RADAR, a disadvantage of the LiDAR sensor is that it cannot simultaneously detect all potential pedestrians in the immediate vicinity.

Yang et al. [28] present RadarNet, a voxel-based sensor fusion sensor fusion method that uses both geometric and dynamic information of RADAR. By fusing LiDAR and RADAR data, pedestrians can be better detected at a greater distance. The Doppler effect is also used to distinguish detected pedestrians from the static environment.

Nobis et al. [29] propose RVF-Net that also uses a voxel structure to process geometric information. Through early feature fusion, geometric and dynamic information is used for fast and precise 3D regional proposals.

3.4   LIDAR Sensor, RADAR Sensor and RGB Camera

Wang et al. [30] combine with HDF-PointNet the direct and lossless processing of 3D point clouds based on PointNet++, the pre-filtering by frustums based on F-PointNet [12] with the temporal information of a RADAR point cloud, similar to RadarNet. The latter also provides the RADAR cross-section (RCS) as a spatial component and the Doppler velocity as a temporal component. Thus, utilizing a multimodal fusion, LiDAR, RADAR and RGB data are spatially and temporally combined.

4   Conclusion and Outlook

In this paper, fundamental as well as state-of-the-art 2D and 3D pedestrian detection methods have been presented. These methods use 2D mono and RGB images from RGB, RGBD and thermal cameras and also 3D point clouds from LiDAR and RADAR sensors. By fusing 2D and 3D as well as 3D-only data, the respective sensor-specific advantages can be combined, and possible disadvantages can be compensated, leading to a more precise pedestrian detection.

The follow-up work will examine the different methods concerning currently valid data protection regulations and their implementation according to the privacy by design principles. In addition, ways will be explored to abstract and securely and reliably transmit personal-critical data over sensor networks.

References

1. OECD: Road Safety Report Germany, International Transport Forum (2021)
2. European Parliament: Regulation (EC) No 78/2009 of the European Parliament and of the Council of 14 January 2009 on the Type-approval of Motor Vehicles with Regard to the Protection of Pedestrians and Other Vulnerable Road Users (2009)
3. You, S., Ji, Y., Liu, S., Mei, C., Yao, X., Feng, Y.: A thermal infrared pedestrian-detection method for edge computing devices (2022). https://doi.org/10.3390/s22176710
4. Geiger, A., Lenz, P., Stiller, C., Urtasun, R.: Vision meets robotics: the KITTI dataset (2013). https://doi.org/10.1177/0278364913491297
5. Redmon, J., Divvala, S.K., Girshick, R.B., Farhadi, A.: You only look once: unified, real-time object detection (2016). https://doi.org/10.1109/CVPR.2016.91
6. Redmon, J., Farhadi, A.: Yolov3: an incremental improvement, CoRR (2018). https://arxiv.org/abs/1804.02767
7. Liu, W., et al.: SSD: single shot multibox detector (2016). https://doi.org/10.1007/978-3-319-46448-0_2
8. Ren, S., He, K., Girshick, R.B., Sun, J.: Faster R-CNN: to- wards real-time object detection with region proposal networks (2015). https://proceedings.neurips.cc/paper/2015/hash/14bfa6bb14875e45bba028a21ed38046-Abstract.html
9. He, K., Gkioxari, G., Dollár, P., Girshick, R.: Mask R-CNN (2017). http://arxiv.org/abs/1703.06870
10. Girshick, R.: Fast R-CNN (2015). https://doi.org/10.1109/ICCV.2015.169
11. Lin, T., Goyal, P., Girshick, R.B., He, K., Dollár, P.: Focal loss for dense object detection (2020). https://doi.org/10.1109/TPAMI.2018.2858826
12. Qi, C.R., Liu, W., Wu, C., Su, H., Guibas, L.J.: Frustum PointNets for 3D object detection from RGB-D data (2018). https://doi.org/10.1109/CVPR.2018.00102
13. Qi, C.R., Su, H., Mo, K., Guibas, L.J.: PointNet: deep learning on point sets for 3D classification and segmentation, CoRR (2016)
14. Zhou, Y., Tuzel, O.: VoxelNet: end-to-end learning for point cloud based 3D object detection (2018). https://doi.org/10.1109/CVPR.2018.00472
15. Chen, X., Ma, H., Wan, J., Li, B., Xia, T.: Multi-view 3D object detection network for autonomous driving (2017). https://doi.org/10.1109/CVPR.2017.691

16. Qi, C.R., Yi, L., Su, H., Guibas, L.J.: Pointnet++: deep hierarchical feature learning on point sets in a metric space, CoRR (2017)
17. Shi, S., Wang, X., Li, H.: PointRCNN: 3D object proposal generation and detection from point cloud (2019). https://doi.org/10.1109/CVPR.2019.00086
18. Lang, A.H., Vora, S., Caesar, H., Zhou, L., Yang, J., Beijbom, O.: PointPillars: fast encoders for object detection from point clouds (2019). https://doi.org/10.1109/CVPR.2019.01298
19. Deng, J., Shi, S., Li, P., Zhou, W., Zhang, Y., Li, H.: Voxel R-CNN: towards high performance voxel-based 3D object detection. CoRR (2020). https://arxiv.org/abs/2012.15712
20. Li, J., et al.: P2VRCNN: point to voxel feature learning for 3D object detection from point clouds (2021). https://doi.org/10.1109/ACCESS.2021.3094562
21. Shi, S., et al.: PV-RCNN: point- voxel feature set abstraction for 3D object detection, pp. 10526–10535 (2020). https://doi.org/10.1109/CVPR42600.2020.01054
22. Shi, S., et al.: PVRCNN++: point-voxel feature set abstraction with local vector representation for 3D object detection, CoRR (2022). https://arxiv.org/abs/2102.00463
23. Wang, K., Zhang, Z.: Point-voxel fusion for multimodal 3D detection (2022). https://doi.org/10.1109/IV51971.2022.9827226
24. Cennamo, A., Kästner, F., Kummert, A.: Towards pedestrian detection in radar point clouds with PointNets, pp. 1–7 (2021). https://doi.org/10.1145/3459066.3459067
25. Wang, Z., Jia, K.: Frustum ConvNet: sliding frustums to aggregate local point-wise features for amodal (2019). https://doi.org/10.1109/IROS40897.2019.8968513
26. Ku, J., Mozifian, M., Lee, J., Harakeh, A., Waslander, S.L.: Joint 3D proposal generation and object detection from view aggregation (2018). https://doi.org/10.1109/IROS.2018.8594049
27. Dimitrievski, M.D., Shopovska, I., Hamme, D.V., Veelaert, P., Philips, W.: Weakly supervised deep learning method for vulnerable road user detection in FMCW radar (2020). https://doi.org/10.1109/ITSC45102.2020.9294399
28. Yang, B., Guo, R., Liang, M., Casas, S., Urtasun, R.: RadarNet: exploiting radar for robust perception of dynamic objects (2020). https://doi.org/10.1007/978-3-030-58523-5_29
29. Nobis, F., Shafiei, E., Karle, P., Betz, J., Lienkamp, M.: Radar voxel fusion for 3D object detection (2021). https://doi.org/10.3390/app11125598
30. Wang, L., Chen, T., Anklam, C., Goldluecke, B.: High dimensional frustum PointNet for 3D object detection from camera, LiDAR, and Radar (2020). https://doi.org/10.1109/IV47402.2020.9304655

Real-Time Aspects of Image Segmentation of Road Markings in Miniature Autonomy

Daniel Riege$^{(\boxtimes)}$, Stephan Pareigis, and Tim Tiedemann

Department Informatik, HAW Hamburg, Berliner Tor 7, 20099 Hamburg, Germany
`daniel.riege@haw-hamburg.de`

Abstract. Image segmentation is used to recognize lanes and markings on the road to provide a basis for autonomous driving. On embedded systems in autonomous miniature vehicles with limited resources, timing aspects have to be considered. Optimisations in hardware design and in the architecture of the artificial neural network speed up the inference such that real-time conditions for autonomous driving can be met. Different architectures for artificial neural networks for segmentation are compared. Inference of the segmentation networks is tested on a TPU, an iPhone, and on a Mac with M1 chip. Experiments show, that inference directly on an iPhone is superior to the other two options.

1   Introduction

In [1] the authors have developed miniature vehicles to find and to study new approaches for autonomous driving in a city-like model environment. Vehicles in the scale 1 : 87 as shown in Fig. 1 have been constructed and equipped with suitable hardware and software. The vehicles use a small camera and a microprocessor to do on-board image based feature extraction of the environment. A TPU (Tensor Processing Unit) [2] is integrated into the hardware of the vehicle to accelerate ML inference. The experimental setup provides a basis for different technological approaches to be applied, investigated, and tested: Several regression networks have been applied, as well as conventional lane recognition with scan lines, various kinds of semantic lane segmentation, imitation learning and reinforcement learning approaches. Some of these approaches combine both image interpretation and steering control, e.g. reinforcement learning, and end-to-end approaches. Other algorithms like lane segmentation require an additional control method for steering to be applied on top of the image recognition. In these latter cases, usually pure pursuit methods are used for steering control (see e.g. [3]).

In [4] the author has trained an artificial neural network to do semantic lane segmentation. Lane markings are detected pixel-wise and interpreted, according to their meaning in traffic control. The recognized lane markings are then subsequently used by other vehicle control algorithms.

In order to enable control for path planning and vehicle guidance of the miniature vehicle, time constraints must be met. The computationally intensive

H. Unger and M. Schaible (Eds.): Real-Time 2022, LNNS 674, pp. 97–107, 2023.
https://doi.org/10.1007/978-3-031-32700-1_11

feature extraction is made possible by specialized hardware adapted to the ML procedures and an optimised architecture of the neural network.

The work compares different artificial neural network architectures for semantic segmentation as well as three different hardware platforms. Section 2 describes the technology of semantic segmentation of lane markings as done in [4]. Section 3 describes three different hardware setups used for the experiments. In Sect. 4 the results of the experiments are presented.

2   Segmentation of Road Markings

In order to plan a route for an autonomous vehicle on the street road markings are essential. To detect road markings in camera images artificial neural networks can be used. These offer advantages in comparison to classical image processing especially in complex environments. Missing road markings, trees and cars blocking the view are examples that make the task of road marking detection complex. To represent the detected markings for further processing, segmented images can be used. Every pixel in the camera image is given a class of road markings. These classes can be derived from road traffic regulations. Possible road marking classes can be dashed or solid lane lines, waiting lines at intersections and more.

2.1   Training Data

The TinyCar CM4 vehicle [5] was used to generate the data sets. The camera of this vehicle contains the required wide angle. With an angle of 120°C horizontally, larger intersection areas can be captured. By remotely controlling the car and not keeping it perfect in the center of the lane, different perspectives were achieved to enhance the robustness of the neural net. Figure 2 shows the area in which the car was set.

Fig. 1. Miniature Autonomous Vehicle. The small camera is looking forward. Steering is realized by a small servomotor (underneath). The computation board contains a Compute Modul CM4 and a Tensor Processing Unit (TPU).

Fig. 2. Top down view onto Kunffingen, a fictional town in the Miniatur Wunderland Hamburg. This environment includes urban streets as well as country roads and an Autobahn. Marked in red are the streets included in the dataset. 708 of the images taken within this town are annotated for training a neural net. (Color figure online)

In order to train a neural network by a supervised learning technique, the images in the dataset need to be annotated. Since segmentation networks are used, images need to be generated with color-coded information. Figure 3 shows an example of the training images and the ground truth segmented images. The classes for these annotations are derived from road traffic regulations.

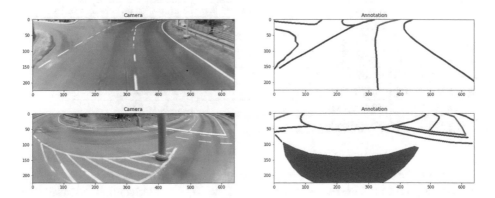

Fig. 3. Two example images from the dataset with annotation (on the right respectively). The camera images are already cropped, so only the lower half of the images are used. The classes define the colors for the annotation. Red = edge of the road, yellow = solid lane line, green = dashed lane line, orange = barrier area, and blue = stop line. (Color figure online)

2.2   Network Architecture UNet

The basic idea behind all presented architectures is that already existing convolutional neural nets (CNN) are used as a basis. These form the backbone of the new UNet models [6]. The CNNs are already pre-trained on the imagenet dataset. Thus, transfer learning is applied here. Since Imagenet is a classification system, these pre-trained networks consist of a convolutional part and fully connected (FC) layers [7–10]. However, segmentation is about classification on the pixel level, which is why the fully connected part is replaced by the newly created decoder. This also has the advantage that the image resolution of the input image can vary. The image size is irrelevant for the convolution kernels of the CNN, only when using FC layers the image size influences the number of weights. The reason for this is that in convolutional layers the weights are located in the convolutional kernels and not, as in fully connected layers, between each possible pixel. Furthermore, all presented architectures use softmax activation in the last layer. In this use case of road markings, each pixel can only belong to one class. Thus, it is a 1 out of n classification on pixel level. Therefore, a softmax activation is used, and not a sigmoid. However, if a pixel does not belong to any class, it must belong to a new background class, otherwise, Softmax will not work. Thus, an artificial background class is created for all classes presented, which contains all unclassified pixels in the annotations.

Table 1. Excerpt of the CNNs that can be used. The number of parameters without FC layer was determined by counting only the parameters up to the last convolution layer. The FC layers are not required since UNet-like architectures are built which use CNN layers on the decoder part as well. Even though VGG16 and VGG19 are the CNNs with the most parameters in total, their CNN layers are relatively small compared to ResNet or InceptionV3. Marked in bold are the two CNNs which are used as the encoder.

Architecture	Number of parameters without FC layer	Number of parameters in total [11]
VGG16 [10]	**14.714.688**	128.357.544
VGG19 [10]	20.024.384	143.667.240
ResNet50V2 [12]	24.564.800	25.613.800
ResNet101V2 [12]	42.626.560	60.380.648
MobileNetV2 [13]	**2.257.984**	3.538.984
InceptionV3 [9]	21.802.784	23.851.784
Xception [7]	20.861.480	22.910.480

Table 1 lists some usable CNNs. The column with number of parameters without FC layer is the relevant one here. Due to the requirement of miniature autonomy and the associated real-time requirements, the two smallest CNNs are used in the further course, thus VGG16 and MobileNetV2.

An example of one of the three UNet architectures created in this paper is shown in Fig. 4. The encoder mainly consists of pooling and convolutional layers.

In the decoder, recurring layers are used. The feature maps are first scaled up by a factor of 2. This is done by linear interpolation. This is followed by a concatenation, which concatenates the feature maps from the upscaling and the respective layer from the encoder (skip connection). This doubles the number of feature maps. In the two subsequent convolutional layers, the number of feature maps is reduced to the number defined for this block. The number of feature maps in the five decoder blocks is always halved by the transposed convolution layer. In total, this architecture consists of 35,901,735 trainable parameters.

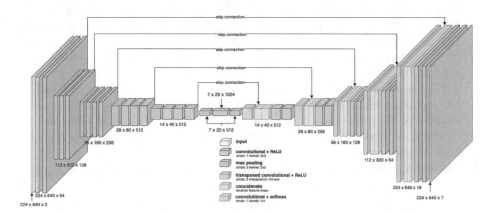

Fig. 4. Representation of the VGGU architecture. The width of the blocks is not proportional to the number of feature maps for display reasons.

The other two architectures are MobileNetU and TinyU, which both have a similar structure as the VGGU architecture. Only the encoder parts are swapped by the MobileNet CNN and the decoder is changed to fit the resolutions of the layers. In total, the MobileNetU has 1.335.111 trainable parameters. The TinyU, compared to MobileNetU, uses a slightly different approach in the decoder. Instead of normal convolutional layers, a depth-wise separable convolution is used. This reduces the number of parameters even more, to 703.109 trainable parameters, but increases the number of operations needed for the inference (Fig. 5).

3   Hardware Setups

A TinyCar CM4 prototype is used as the experimental platform [5]. The TinyCar CM4, as seen in Fig. 1, is a 1:87 scale road vehicle with the necessary peripherals to enable fully autonomous driving. It uses a Raspberry Pi Compute Module 4 (CM4) as the main computing unit and a Google Coral TPU (Tensor Processing Unit) as a machine learning accelerator. A front-facing ultra-wide-angle camera with a viewing angle of approx. 120° is used for the recognition of the immediate environment. Linux and ROS are used as the operating system.

Three different hardware setups have been tested and are compared. Figure 6 shows an overview from a software perspective.

1. The TinyCar CM4 with the Google Coral TPU is the first setup used to perform latency measurements of the three network architectures. In order to run the trained models on the TPUs, they need to be converted into a Tensorflow Lite model which is 8-bit quantized. With this 8-bit quantization from 32-bit weights, the accuracy of the segmentations suffers. See Fig. 6a.
2. In the second setup used, the TinyCar CM4 sends the preprocessed images of the camera via a TCP stream to a Mac with M1 chip. Using the M1, the inferences of the trained models are calculated on the Apple Neural Engine (ANE). All operations of the three network architectures are fully supported by the Apple Neural Engine. In order to utilize the models for the ANE, they need to be converted into a CoreML model. In contrary to the Google Coral TPU, the ANE uses 32-bit weights. Therefore there is no loss in accuracy of the segmentations when using the ANE. For the TCP stream, the images are sent 24-bit RGB encoded. The connection between the TinyCar CM4 and the router is done over WiFi in a 5 GHz network. The connection between the Mac and the router is wired. See Fig. 6b.
3. The third setup uses the ANE as well, but on an iPhone with an A13 Bionic chip. Since it is theoretically possible to fit the needed components of the iPhone in a 1:87 car, no TCP stream is used. Like the first setup, the inference is done on the device with the camera attached. The A13 Bionic chip is one year older than the M1, therefore the ANE is in an older version on the iPhone. As for the M1, the trained models need to be converted into a CoreML model. See Fig. 6c.

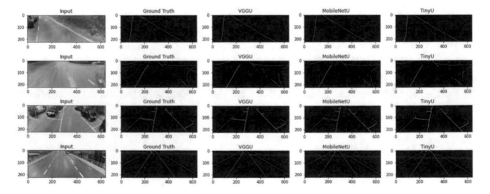

Fig. 5. Example segmentations with all three neural network architectures on the validation set. On the left side, the input is shown, followed by the hand-labeled ground truth and the predictions by the neural net.

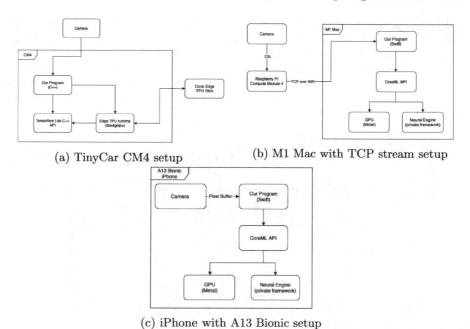

(a) TinyCar CM4 setup

(b) M1 Mac with TCP stream setup

(c) iPhone with A13 Bionic setup

Fig. 6. Visualization of the three hardware setups from a software perspective. (a) shows the setup where the TPU on the board of the vehicle is used for ML-inference. (b) shows the setup where the image data is streamed to the M1 Mac. (c) shows the setup where images are captured with the iPhone camera and ML-inference is calculated directly on the iPhone.

4 Latency Measurements

Latency measurements are made for all possible configurations of hardware setups and neural network architectures. The results are shown in Table 2. For the TinyCar CM4 and iPhone setup, the latency measurements are the inference time of the neural network. This time only depends on the architecture, resolution and hardware used. Beginning with a complete frame in the buffer and ends when the segmented frame lies in another buffer. In the Mac setup, however, which gets the camera frames via a TCP stream from the TinyCar CM4, the total latency besides the inference time includes the latency of the TCP stream. The TCP stream latency does not depend on the network architecture but on the resolution, the network setup and hardware used. As in the two setups before, the measurement begins with a camera frame in the buffer on the TinyCar and ends with a segmented frame in a buffer on the Mac. For every pair 100 latency measurements are made in serial and the average is taken.

As seen in Table 2 the direct inference on the TinyCar CM4 with the TPU has the highest latencies with a maximum of 5.852 s. With smaller architectures and lower resolutions, this time decreases but never reaches the latencies of the setups which are using an Apple Neural Engine. The Mac and iPhone setup

Table 2. Latency measurements in average for different setups. Every setup was tested with each architecture resolution pair. The latency for the TinyCar CM4 and iPhone setup begins with a camera frame event and ends with the segmentation result. These latencies are therefore the inference time. For the TCP stream to Mac setup, the latency measurement begins with a camera frame event on the TinyCar CM4 and ends with the segmentation result on the Mac. Therefore these latencies combine the inference time with TCP stream latency.

Architecture	Resolution	Latency [ms]		
		TinyCar CM4	TCP stream to Mac[1] TCP + Inference	iPhone[2]
VGGU	640×224	5852	41 + 61 = 102	73
	320×96	1295	12 + 7 = 19	22
	224×96	1078	8 + 6 = 14	18
MobileNetU	640×224	871	41 + 15 = 56	41
	320×112	79	12 + 3 = 15	14
	224×96	27	8 + 2 = 10	10
TinyU	640×224	1160	41 + 14 = 55	39
	320×112	58	12 + 2 = 14	13
	224×96	26	8 + 2 = 10	9

[1] with M1, [2] with A13 Bionic

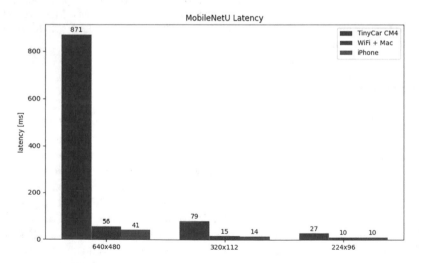

Fig. 7. Latency measurements in average for the MobileNetU Architecture in 3 different resolutions for every setup. It is averaged on 100 values. The WiFi+Mac has lower resolutions and about the same latency as the iPhone. The TinyCar CM4 in contrast is 15.5 times slower in 640×480, 5.2 times slower in 320×112, and 2.7 times slower in 224×96 than the WiFi+Mac setup.

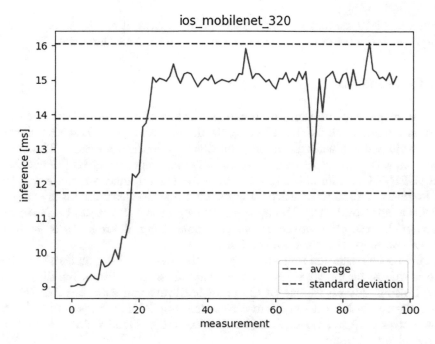

Fig. 8. Inference measurement overtime on the iPhone with A13 Bionic chip. 100 measurements are recorded. The red line represents the average and the green the standard deviation in both directions. In the beginning, inference on one image takes about 10 ms; over time, this increases to about 15 ms. This can be seen in every architecture on the iPhone. (Color figure online)

show similar latencies, even though the inference on the Mac with M1 is faster. The TCP stream costs a lot of time for the Mac setup.

Figure 7 shows the latencies in the three different resolutions for the MobileNetU architecture in a bar chart. This chart gives an excellent visual representation of the factor by which the Apple Neural Engine setups are faster compared to the TPU. In the lowest resolution, however, the difference is not that big anymore. Lowering the resolution of the images can therefore improve the performance significantly, especially on the TPU.

Figure 8 shows the latency (inference time) of the MobileNetU architecture in the 320×112 resolution throughout 100 measurements on the iPhone. The first 20 measurements show a latency of around 10 ms. After these 20 measurements, the latency increases to about 15 ms. This can be seen with every network architecture and every resolution on the iPhone. An analysis shows that after the 20 measurements the program switches from the performance core to the efficiency core. These are cores named by apple for the two cores, which run at a higher frequency (performance), and four cores, which run at a lower frequency (efficiency), on the A13 Bionic chip. This core switching might be due to increasing heat on the chip, resulting in using a lower frequency to produce less heat.

However, this needs further investigation. Nonetheless, it can be shown that the iPhone with the performance cores can achieve even lower latencies.

5   Conclusion

Several image segmentation neural network architectures were tested and evaluated on a hand-labeled dataset in a miniature environment. These architectures were evaluated against their latency on different hardware setups. The experiments show that inference directly on an iPhone is superior, given the instability of a WiFi TCP stream in hardware setup 2. In general, however, the Apple Neural Engine is a good alternative for embedded machine learning tasks compared to the Google Coral TPU. To improve latency on the TPU, it can be shown that lowering the resolution of the images can make a big difference in order to use this hardware for real-time applications.

Further investigation is needed to prove the core switching on the A13 Bionic chip is due to high heat when running inferences. In addition, building a prototype using only an iPhone as the main hardware can show the feasibility of using the Apple Neural Engine in a real-time application. This could not only show better performance but also be more resource friendly since iPhones can be recycled first-hand.

References

1. Tiedemann, T., Schwalb, L., Kasten, M., Grotkasten, R., Pareigis, S.: Miniature autonomy as means to find new approaches in reliable autonomous driving AI method design. Front. Neurorobot. **16**, 846355 (2022)
2. Google, "Coral," (2022). https://coral.ai
3. Nikolov, I.: Verfahren zur Fahrbahnverfolgung eines autonomen Fahrzeugs mittels Pure Pursuit und Follow-the-carrot. Thesis, University of Applied Sciences Hamburg, B.S (2009)
4. Riege, D.: "Segmentierung von Straßenmarkierungen durch maschinelles Lernen für die Miniaturautonomie," B.Sc. Thesis, HAW Hamburg, Department Informatik, Berliner Tor 7, 12 2021
5. Kasten, M.: "Hardwareplattform für Autonome Straßenfahrzeuge im Maßstab 1:87," B.Sc. Thesis, HAW Hamburg, Department Informatik, Berliner Tor 7, 10 2021
6. Ronneberger, O., Fischer, P., Brox, T.: U-Net: convolutional networks for biomedical image segmentation, May 2015. arXiv:1505.04597
7. Chollet, F.: Xception: deep learning with Depthwise separable convolutions, April 2017. arXiv:1610.02357
8. He, K., Zhang, X., Ren, S., Sun, J.: Deep residual learning for image recognition, December 2015. arXiv:1512.03385
9. Szegedy, C., Vanhoucke, V., Ioffe, S., Shlens, J., Wojna,Z.: Rethinking the inception architecture for computer vision, December 2015. arXiv:1512.00567
10. Simonyan, K., Zisserman, A.: Very deep convolutional networks for large-scale image recognition, April 2015. arXiv:1409.1556

11. Team, K.: Keras documentation: Keras Applications. https://keras.io/api/applications/
12. He, K., Zhang, X., Ren, S., Sun, J.: Identity mappings in deep residual networks, July 2016. arXiv:1603.05027
13. Sandler, M., Howard, A., Zhu, M., Zhmoginov, A., Chen, L.-C.: MobileNetV2: inverted residuals and linear bottlenecks, March 2019. arXiv:1801.04381

Real-Time Audio Classification to Determine the Article of a German Noun

Dena Zaiss[✉], Iman Baghernejad Monavar Gilani, and Dietmar Tutsch

University of Wuppertal, Rainer -Gruenter -Str. 21, 42119 Wuppertal, Germany
zaiss@uni-wuppertal.de

Abstract. The determination of an article for gender assignment to a noun is different in every language and depends on the certain rules as well as the context of a word in a text. The rules given in the German language help with the article assignment, but are not always consistent and meaningful for all words. In the human brain, this assignment can take place intuitively and in real-time, if it is trained over many years. This work is concerned with verifying whether a convolutional neural network (CNN) can match an article to a word the same way as the human brain, using methods of audio classification in deep learning. The chosen attribute to train the model is the sound of the word in this work, because the sound plays a big role in the article determination.

Keywords: Convolutional neural network · Deep learning · Audio classification · Article assignment

1 Introduction

In any language, there are many foreign words that integrate into the language over time. In the German language, this is associated with a challenge of article determination. For native German speakers, this is much easier than for those who have learned German as a foreign language. When a new foreign word is used in German, many Germans will instinctively use the same article. How this assignment is done depends on a few characteristics such as the sound of the word, the meaning, and certain grammatical rules [1]. Some rules for article recognition in the German language are known. For example, if a word gets the ending "-ent", the word is 100% masculine and if the ending is "-e", the word is 90% feminine. These rules do not apply to foreign language words, because the word structure is quite different, but can be used for some properties like sound of the word, e.g. for the words ending with "-e" or "-o". These rules can be thought but not every one know them by heart. A native German speaker classifies the words intuitively. For a person who learns German as a foreign language there are some other aspects like to memorize the article or to guess it while speaking. All of these must happen in real-time because making sentences happens in real-time. The real-time we consider here is the soft kind, since there is no specific time limit in general. We have set the time limit to one second in this project.

© The Author(s), under exclusive license to Springer Nature Switzerland AG 2023
H. Unger and M. Schaible (Eds.): Real-Time 2022, LNNS 674, pp. 108–114, 2023.
https://doi.org/10.1007/978-3-031-32700-1_12

Based on the research in sound and audio classification, the first step is to create a dataset of audio files and convert them to an image format that is readable to our neural network [2]. There are several libraries, which make this possible such as Torchaudio [3] and Librosa [4] which is used in this work. After converting the audio files, they must be labled in three different classes, "Die" for Femininum, "Der" for Maskulinum and "Das" for Neutrum and fed to the network.

2   Implementation

2.1   CNN

Since CNNs are built on the mathematical principle of convolution and this approach is particularly suitable for image- and audio recognition, they have been chosen for this work. The result of the convolution is the reduction of the data and thereby the increase of the calculation speed without decreasing the performance [5].

2.2   Structure

The structure of our Convolutional Neural Network is as follows:

At first we started with three convolutional layers which we changed later for optimization purposes, two pooling layers, one flatten layer and two dense layers. Our dataset class uses two inputs: saved images created from the audio files and CSV-files which include the name of the images and the label of each of them. This class transforms the images in 28×28 pixels.

The optimization algorithm which is used in this work is SGD (Stochastic Gradient Descent).

2.3   Data Creation

The first step to create a dataset is to record the sound of the words. The second step is to convert the audio files from time domain to frequency domain and to convert them to an image. In Fig. 1 above on the left side an audio file is shown in time domain and on the right side the same audio file in frequency domain. Our dataset has 3308 objects with 1294 objects labled with "Der", 1359 with "Die" and 655 with "Das". The Test set is 6.5% of the whole data and the validation set is 20% of the training set.

2.4   Overfitting Prevention

Overfitting can occur when the model is trained very well on the training data but works poorly on new data. To prevent overfitting, two methods will be reviewed in this work. K-FOLD cross validation [6] and Early Stopping [7].

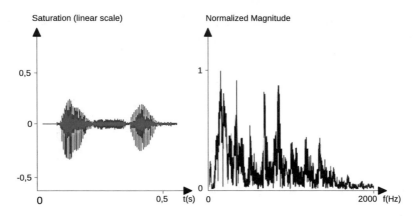

Fig. 1. Voice waveform over time (left) converted to audio frequency spectrum (right)

K-Fold Cross Validation. The data set is randomly shuffled and split in k groups from which every individual group will be chosen as the test set once and the others as the training set. The model will be trained k times and tested with a new independent data set.

Early Stopping. Early Stopping stops the training, whenever there is a chance of overfitting. It monitors the model performance and stops the training before overfitting occurs. For this purpose a number of iterations (epochs) are chosen in which there are no performance improvement but error (loss) increase.

3 Results

The first results show an accuracy of 45–49% for the classified objects in real-time. Using methods of handling imbalanced dataset and modifying the dataset creation, we were able to create improved results.

Results with K-Fold Cross Validation. In training with K-Fold Cross Validation, our data set is divided into 5 groups. One of the groups is used as the test set and the rest as the training set. Since there are 5 interdependent test sets, the model is trained 5 times and the results are compared. The results (Table 1) show an accuracy of maximum 49.17%.

Results with Early Stopping. In training with Early Stopping we have set the steps in which the loss (error) is less than the last minimum loss to 10. Figure 2 shows that the training stops after 35 iterations and the minimum loss is at the 25th one.

Table 1. Results with K-fold cross validation

FOLD 0	Average loss: 1.10089, Accuracy: 45.17 %
FOLD 1	Average loss: 1.03677, Accuracy: 46.68 %
FOLD 2	Average loss: 0.97006, Accuracy: 49.17 %
FOLD 3	Average loss: 1.09230, Accuracy: 46.14 %
FOLD 4	Average loss: 0.99008, Accuracy: 46.75 %

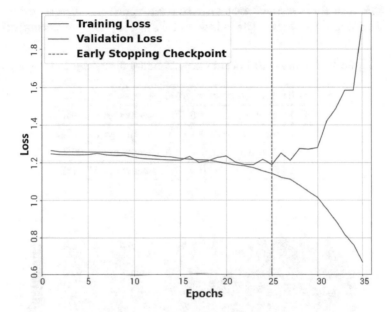

Fig. 2. Results with Early Stopping

3.1 Optimization Methods

MFCC. To optimize the results, a few changes have been introduced to the network. The first is to use the MFCC method to analyse the images in time domain. The basis for MFCC (Mel Frequency Cepstral Coefficients) is the linear modelling of speech generation. A periodic excitation signal generated by vocal cords is shaped by a linear filter like the mouth. Since the word itself and not its pitch is of interest in this work, it is important to separate the impulse response of the linear filter from the periodic excitation signal which is done by using the MFCC method [8].

Using the ADAM-Function. Adam (adaptive moment estimation) is an optimization algorithm which is used here as a replacement for SGD. It is easy to implement and it is computationally more efficient on noisy data than the SGD. [9].

Deepening. For optimization one more convolutional layer is added to deepen the network and to optimize the results.

3.2   Accuracy Analysis Using Confusion Matrix

After using optimizing methods, we were able to increase the accuracy. Table 2 shows the accuracy of K-Fold Cross Validation after optimizing the results. Figure 3 shows the confusion matrix for the first Fold (Fold 0) as an example. In this case there are 200 test objects from which 44 objects are classified in the "Der"-class 35 in the "Die"-class and 24 in the "Das"-class correctly.

Table 2. Results with K-Fold Cross Validation after optimization

FOLD 0	Average loss: 1.03296, Accuracy: 51.50%
FOLD 1	Average loss: 1.06094, Accuracy: 50.50%
FOLD 2	Average loss: 0.96576, Accuracy: 52.00%
FOLD 3	Average loss: 1.02503, Accuracy: 53.00%
FOLD 4	Average loss: 0.98233, Accuracy: 51.01%

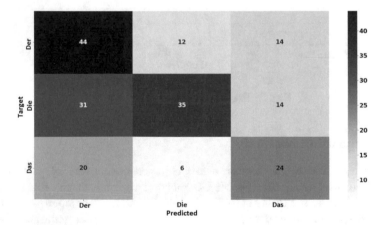

Fig. 3. Accuracy calculation for Fold 0 with confusion matrix

With Early Stopping there is both an increase in accuracy and an optimization of the training time. Figure 4 shows that the minimum is at the 13th iteration step (Epoch). Figure 5 shows the confusion matrix and an increase of the accuracy from 48% to 57%.

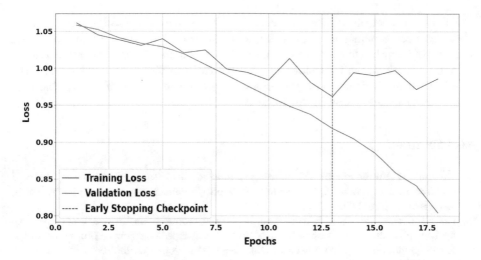

Fig. 4. Results with Early Stopping after optimization

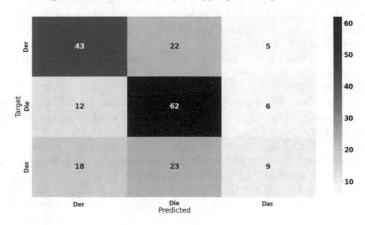

Fig. 5. Accuracy calculation for Early Stopping with confusion matrix

3.3   Time Optimization

Using the optimization methods we are able to reduce the training time from 90 to 20 min. As for the classification time, the minimum run time for classifying a single object is 6 ms and the maximum 14 ms. That means with the average run time of 10 ms we have a reliable classification time under one second.

4   Conclusion and Outlook

The results show that our neural network delivers results that are better than just randomly guessing. The accuracy of the results is between 50% and 57% and the time limit of one second is respected.

To train a neural network to recognize the articles of a foreign noun in real-time, one has to consider many other aspects such as meaning of the word and context of it in the text but also statistical components. To increase the accuracy of the network we will consider these features in addition to the sound of the word in future projects. Also we hope to improve the results, if we will increase the number of the words in our dataset.

References

1. Genus bei Fremdwörtern-Variantengrammatik des Standarddeutschen, http:// mediawiki.ids-mannheim.de/VarGra/index.php/Genus_bei_   Fremdwörtern, Accessed 31 May 2022
2. Doshi, K.: Audio deep learning made simple: sound classification, step-by-step by Ketan Doshi towards data science. https://towardsdatascience.com/audio-deep-learning-made-simple-sound-classification-step-by-step-cebc936bbe5. Accessed 15 May 2022
3. Yang, Y. -Y., et al.: Torchaudio: building blocks for audio and speech processing. In: ICASSP 2022–2022 IEEE International Conference on Acoustics, Speech and Signal Processing (ICASSP), pp. 6982–6986 (2022). https://doi.org/10.1109/ICASSP43922.2022.9747236
4. McFee, B., et al.: librosa: audio and music signal analysis in python. In: Proceedings of the 14th python in Science Conference, pp. 18–25 (2015). https://doi.org/10.5281/zenodo.4792298
5. [at] REDAKTION. Convolutional Neural Networks [at] Blog. https://www.alexanderthamm.com/de/blog/convolutional-neural-networks-am-beispiel-der-revolution-der-computervision/. Accessed 31 May 2022
6. Allibhai, E.: Hold-out vs. cross-validation in machine learning—by Eijaz Allibhai—medium,   https://medium.com/@eijaz/holdout-vs-cross-validation-in-machine-learning-7637112d3f8f. Accessed 10 Apr 2022
7. How to implement early stopping in PyTorch - Quora, https://www.quora.com/How-can-I-implement-early-stopping-in-PyTorch. Accessed 15 Mar 2022
8. Logan, B.: Mel Frequency Ceptral Coefficient for Music Modeling. Cambridge Research Laboratory, Cambridge (2000)
9. Brownlee, J.: Gentle Introduction to the Adam Optimization Algorithm for Deep Learning. Machine Learning Mastery, San Juan, Blog (2021)

Live GNSS Tracking of Search and Rescue Dogs with LoRa

Magdalena Thomeczek[(✉)]

University of Applied Sciences Landshut, Am Lurzenhof 1, 84036 Landshut, Germany
`realtime@losara.de`

Abstract. Search and Rescue Dog training is a complex and time-consuming topic. Real-Time visualisation of area search dogs could support training and missions. This was acheived with cost effective IoT hardware and a smartphone app. This article presents work from the author's Master Thesis highlighting the Real-Time requirements of the use-case and their implementation.

Keywords: SAR · Dog · Real-Time · LoRa · GNSS · GPS · App · Android

1 Introduction

Dogs are uniquely equipped to support humans in various ways, one of which is Search and Rescue. With their superhuman sense of smell they can quickly locate missing persons in large areas that would otherwise need to be searched by many humans manually. With the support of the Johanniter Rettungshundestaffel Landshut a proof of concept for a live visualisation of Search and Rescue dog tracks was elaborated and evaluated.

For a dog to become a reliable Search and Rescue Dog at least two years of training are required. After passing a test such a dog can be used in missions, but the training has to continue so the dog stays in shape. Just like in Machine Learning the more situations the dog has experienced, the more likely it is that it can generalize and correctly find and indicate a missing person. Training and missions are very time consuming and additional training for the handler is required. Despite this Search and Rescue dog work is mostly done by volunteers.

In area search the dog runs loose around the handler and searches for human scent. Area search is usually used to search large sight obstructed areas e.g. forests or rough terrain. The dogs often go out of sight and while they are equipped with a bell for audible feedback on their position, the exact paths a dog takes are not exactly traceable on the spot. This complicates training. It can be crucial to know if a dog found a subject in the training and if not to work out why.

How easily a person can be found depends greatly on the conditions on the spot like slopes, time of day, weather and wind. These effects impact the air flow and thus the scent distribution. Obstacles like vegetation or structures can hinder the scent distribution and impede discovery [1]. Where a dog has to go to find humans includes these factors, but is not limited to them. While knowing

H. Unger and M. Schaible (Eds.): Real-Time 2022, LNNS 674, pp. 115–119, 2023.
https://doi.org/10.1007/978-3-031-32700-1_13

where a dog has been is highly beneficial, it is only useful on site where all these factors can be analysed and taken into account.

In a mission the dog handler has to tactically guide the dog through a defined area. The handler has to choose a tactic that covers all possible locations a missing person could be. As this tactic depends on all the above-mentioned influences the dog handler has to decide on the spot, when to search roughly or when to search more fine-grained. In the end the search area has to be declared fully searched, with confidence, because this could result in this area not being searched again.

Visual feedback on the covered area by visualisation of the dog's path can assist in that decision. The track of the dog is often logged in missions with GPS loggers as a validation in case the missing person is later found in this area. However, due to the dependence on the on site circumstances this information is only useful live and on site.

In summary, live visualisation could greatly support training and missions of Search and Rescue dogs. Current solutions do not offer local transmission, a suitable update frequency (best is 2 s) or are expensive (> 500 €) [2]. To ensure low cost the idea evolved to combine a GNSS receiver with a sender using commonly available radio technologies to transfer the location locally to a smartphone.

2   Methods

With the emergence of the internet of things new ways of wireless transmission were required to connect remotely located sensors. One of these is the radio modulation technique Long Range (LoRa) [3]. As the name suggests it is used for long range transmission of data. It can be used bidirectionally and directly without the need for deployed hardware. This makes it uniquely suitable for this use-case in comparison to alternatives [2].

LoRa uses commonly available frequencies (in Europe 868 MHz) [4], which are subject to duty cycle limits restricting the Time on Air (ToA) [5]. The duty cycle limits restrict how often a location update can be send and thus directly infuences the usability for certain rescue dog tracking. An update frequency of at least once per second is desireable. A dog running up to 40 km/h makes around 11 m/s. While a dog is seldom this fast in a forest an update frequency of every 5 s like the most notable current alternative uses [6] is not precise enough. The duty cycle limits must allow the desired update frequency. To ensure this, the ToA for different message lengths was calculated and measured with a Nooelec NESDR SMArt v4 RTL-SDR Bundle with the UHF Antenna and CubicSDR as recording tool at 868 MHz, with a 2.048 MHz sample rate.

The range of LoRa depends on certain factors, such as bandwidth or spreading factor. A trade-off has to be made between Time on Air and range [7]. To ensure a sufficient update frequency, range tests were conducted with an optimization for short Time on Air and thus worst range. The last/first coordinate received was compared to the coordinate of the receiver while leaving/approaching the receiver location in different settings in a typical mission area. Furthermore the whole functional set-up was used in practical tests during Search and Rescue dog training to ensure the range is sufficient in practice.

The hardware used was chosen for rapid prototyping. The developement boards by Pycom are called LoPy4 and Pytrack2.0 and use and ESP32 micro-controller [8,9]. They can be programmed with MicroPython and come with short examples for use of LoRa and GNSS [10]. Since the internal GPS antenna that comes with the Pytrack is very small a passive and an active antenna were used as well and their suitability compared.

The app was created according to elaborated must have requirements and the set-up tested in trainings. The resulting dog tracks were compared to location loggers previously used by the dog unit.

The sender was attached in a waterproof plastic container and attached to the dog vest marking the dog as a search and rescue dog. To ensure that the sender can be used in all weather conditions and if a dog goes into water, impermeability is crucial. The sender should ideally be light and small and fall off if a dog gets stuck in underwood or the like, for the dog's safety.

Real-time requirements were deemed as follows: Real-time GNSS positioning, direct local data transfer every second with a live visualisation after a maximum of a few seconds.

3   Results

After simple range tests with the example code [10] in urban area (see [2]) deemed the technology potentially suitable for the case, further work was conducted. A basic prototype consisting of two devices and a smartphone was created and tested in dog training, by giving the sender to a handler.

The results showed that the system can be used to show a track with several seconds delay (up to 30 s). It yielded interesting results as it showed clearly visible differences between the search styles of different handler/dog teams (see [2]). Different search styles are normal and can be used tactically. It emerged that visualisation documents search styles and training progress and that the range was sufficient so far.

3.1   Usability of LoRa

Results showed that calculations of ToA could be confirmed through measurements. Only frequencies with a 10% duty cycle can be used with the desired update frequency. Further optimisation of the payload or by velocity dependant sending of messages (more while fast, less while slow) could make other frequencies usable as well.

The range was sufficient yielding ranges of 218 m, 322 m and 463 m with forest in-between, on a forest road and out in the open respectively. This is not ideal but testing in practice showed that range was sufficient in most situations. Range could be extended further with optimisation of ToA. With a payload of 13 Bytes a ToA of 19 ms was calculated and measured and frequency bands of

10% duty cycle the range could used for up to 100 ms, showing there is room for enhancing the range.

3.2 GNSS Reception

A comparison of the internal antenna and a passive antenna yielded a statistically significant increase in satellites in view ($p<0,05$) with the passive antenna. Visual differences, in comparison with various other location tracking devices (Smartwatch, Smartphone, Garmin 65S) showed that the internal antenna track differed notably from other tracks and the passive antenna was better but not ideal.

For the final set-up a an active ceramic antenna was used and compared with smartphone position and previously used GPS loggers. The loggers used were not very accurate most likely due to their age. The set-up proved more accurate and highly effective in training - showing even difficult to track movements such as a dog entering different small kennels (around $1.5 \, m \times 3 \, m$) in a building. This was better than before and exceeded the expectations.

3.3 Real-Time Requirements

Tests in practice showed that visualisation with a delay of more than three seconds were difficult to compare with the actual path of the dog by watching the dog and visualisation, however this is only a usability and potentially user trust problem. In trainings or missions the path is not necessarily monitored live but only at certain points when taking a break. If the visualisation takes a few seconds to catch up the concept would still be useful.

Hardware running with less overhead (e.g. C instead of MicroPython) could be faster and allow better optimisation and less delay.

3.4 Supporting the Training

Most important for this kind of system is if it is useful in the field. The system was tested in real search and rescue dog training several times at different stages and the results were highly promising. The different search styles were easily visible from beginning on and could be used to tactically assign areas in missions.

In one search during training a suspicious behaviour went unnoticed and the track later looked like the dog was very close to a subject. Considering the track visualisation only, it could look like the dog was at the subject and did not alert - a mistake that has to be known and training adjusted accordingly. However, later in the search the subject was found and the reason for the strange behaviour and the missing alert clear: There was dense vegetation which allowed for a bit of scent to pass through, but was too difficult for the dog to work through [2].

If the dog handler had assessed the behaviour correctly, they could have helped the dog find a way around the vegetation and find the subject much earlier in the search. This might not be completely relevant in training, in missions however, time is crucial. This was a lesson learned for the dog handler and therefore a first real success for this system. Additionally, remembering the behaviour and putting it together with the track and position of the dog is much easier during or right after the search, again illustrating the importance of meeting the real-time requirements.

4   Conclusion

As outlined in the introduction the visualisation develops its full potential only when meeting real-time requirements. The thesis has shown that affordable hardware together with an app can greatly support search and rescue dog work. Further work is needed to make the concept widely available for dog units. Hardware that can be used by non-computer scientists is required, more functions in the app and thorough stability testing. The best trade-off between choice of frequencies, ToA, payload length and range should be determined or alternatively other ways of local communications could be researched and used with the app.

Search and Rescue dog training is not well researched and often done by volunteers with limited funds. More research for technologies or in general could help save lives.

Acknowledgments. Many thanks to the Johanniter Rettungshundestaffel Landshut.

References

1. Wegmann, A.: Such und Hilf! Ein Handbuch für Ausbildung und Einsatz des Rettungshundes. Kynos, Nerdlen (2021)
2. Thomeczek, M.: Live Tracking of Search and Rescue Dogs with LoRa (2022)
3. Semtech: LoRa (2022). https://www.semtech.com/lora/what-is-lora. Accessed 2 May 2022
4. Adroher, A.: 5.2.2.4 LoRa · Pycom Documentation (2022). https://alepycom. gitbooks.io/pycom-documentation/content/chapter/firmwareapi/pycom/ network/lora.html. Accessed 10 June 2022
5. Bundesnetzagentur: Allgemeinzuteilung von Frequenzen zur Nutzung durch Funkanwendungen geringer Reichweite, SRD, Vfg. 133/2019 (2020)
6. Garmin: GPS-Hundehalsbänder, Hundeortungsgeräte, Garmin (2022). https:// www.garmin.com/de-DE/c/outdoor-recreation/sporting-dog-tracking-training-devices. Accessed 6 July 2022
7. The Things Network: Airtime Calculator (2022).https://www.thethingsnetwork. org/airtime-calculator. Accessed 06 June 2022
8. Pycom: LoPy 4 (2022). https://docs.pycom.io/datasheets/development/lopy4/. Accessed 15 June 2022
9. Pycom: Pytrack 2.0X (2022). https://docs.pycom.io/datasheets/expansionboards/ pytrack2/. Accessed 15 June 2022
10. Pycom Docs: LoPy to LoPy (2022). https://docs.pycom.io/tutorials/networks/ lora/module-module/. Accessed 15 June 2022

AutSys

Artificial Intelligence in Autonomous Systems. A Collection of Projects in Six Problem Classes

Stephan Pareigis[✉], Tim Tiedemann, Nils Schönherr, Denisz Mihajlov,
Eric Denecke, Justin Tran, Sven Koch, Awab Abdelkarim,
and Maximilian Mang

Department Computer Science, University of Applied Sciences Hamburg,
Berliner Tor 7, 20099 Hamburg, Germany
`stephan.pareigis@haw-hamburg.de`

Abstract. The paper presents a collection of projects on autonomous mobile systems with a focus on artificial intelligence technologies. Basic technologies necessary for autonomous mobile behaviour are described. Current methods and application fields of SLAM, feature extraction based on various sensory data, working with simulations, and gripping with robot manipulators will be discussed. Some ideas on solutions for problems in real world applications will be presented. The area of application is a four wheeled robotic platform with an integrated 6-DOF manipulator. Focus will be set on machine learning methods, in particular object detection and reinforcement learning. The paper gives an overview of required technologies for autonomous robotic systems and current state of the art methods. Some aspects and problems arising in applications will be discussed in more detail. The projects have been carried out in student projects at the autosys research lab at the Department Computer Science of University of Applied Sciences Hamburg [1].

Keywords: autonomous systems · robotics · SLAM · object detection · robot manipulation · point cloud registration

1 Introduction

Simple autonomous or semi-autonomous mobile robots can either be created with little effort or bought ready-made. There are many manufacturers of mobile robot platforms that already have ready-made autonomous behavior: lawn mower robots, robots for training in schools and universities, robots for industrial applications, etc. An autonomous mobile system (AMS) usually consists of a driving platform. Various drive options are conceivable for the driving platform, which are discussed in more detail in Sect. 2. In addition to that some sensors are necessary, which are responsible for detecting the environment. A manipulator can be attached to the AMS for interaction with the environment as an option. Finally, there is a computer platform that runs the processing and controlling software.

© The Author(s), under exclusive license to Springer Nature Switzerland AG 2023
H. Unger and M. Schaible (Eds.): Real-Time 2022, LNNS 674, pp. 123–145, 2023.
https://doi.org/10.1007/978-3-031-32700-1_14

In this work, basic technologies which are necessary to build an autonomous system are introduced. Based on a stable robotic platform that has some basic capabilities of environment sensing, mapping and trajectory planning, more complex autonomous capabilities are then added. Artificial intelligence methods are used for this. The focus of the work is on the application of these methods in autonomous mobile systems.

In Sect. 2 the essential components of a stable autonomous robot platform are presented. A hardware platform with appropriate sensors and software is described that enables basic autonomous behavior. Different drive forms and suitable sensors and algorithms are discussed.

Section 3 explains how a map of an outdoor environment can be created. A functional experimental setup is presented, which leads to good results. Typical problems that can lead to failure are mentioned.

Section 4 describes how objects can be scanned by the autonomous system in order to make them known to the system. Registration of point clouds and surface reconstruction are the core of this chapter.

Section 5 explains several capabilities of the autonomous mobile system in buildings: driving elevators, recognizing doors and opening doors.

In Sect. 6 two suggestions for the interaction of humans with autonomous systems are made. One is a passive system based on visual effects. The other system is an emergency stop device based on recognizing voice commands.

In Sect. 7 a positioning system is presented, which can be used in urban areas where GPS reception is too imprecise.

2   Basic Autonomous Behaviour

Task. Equip a mobile robot with basic autonomous behaviour based on existing technology.

Discussion. We will begin with a discussion on different drive technologies in autonomous platforms. There are different aspects to consider when choosing a platform. An important aspect is the odometry. The position, velocity and acceleration of the vehicle is measured in many cases using wheel encoders.

The first images in Fig. 1 and Fig. 2 respectively show robots with a differential drive and a support wheel. Odometry is easy in these cases as there is almost no slip on the floor. The same holds true for vehicles with an Ackermann steering as shown in the top row of Fig. 1 images 2, 3 and 4 when the steering angle is known.

Problems with odometry which is based solely on wheel encoders arise with vehicles that use a (differential) skid-steering drive as shown in the bottom row of Fig. 1 images 1 and 2, as well in Fig. 2 (images in the middle and right). While the steering concept still is differential like described before, a skid-steering drive needs wheels or the chain to slip while turning. Because of different surface or soil conditions the slip is unpredictable and correct turning angles of the vehicle cannot be predicted based solely on wheel encoders. Furthermore the pivot point of the rotation can vary also. These vehicles often use IMUs (inertial measurement units) to assist in calculating the odometry.

When a ship is used as an autonomous platform (see right image in the bottom row of Fig. 1), visual odometry may be used if landmarks are available. In the use case of a model ship inside a harbour as shown in the image, infra-red distance sensors may be used for positioning the ship inside a given map of the harbour.

There are also many more drive technologies, e.g. Mecanum Wheels, which allow the robot to drive in arbitrary directions on a plane (holonomous). While odometry can be calculated, it is dependent on surface contact of all wheels. So this drive technology is working best on flat surfaces.

Fig. 1. Multiple autonomous platforms which have been constructed in the autosys research lab [1]. Top row from left to right: The first image shows a small differential drive platform controlled by a small microcontroller which is used in cooperative scenarios. In the next image are two platforms which drive on street markings, a 1:10 scale and a 1:63 scale. The next two images show a 1:87 scale autonomous platform which also drives on streets and the 1:63 scale platform in detail. These two tiny vehicles are also used as basic platforms in our *miniature autonomy* approach, see [2]. Bottom row: Left two images show skid-steering drive platforms. To the right there is a ship equipped with infrared distance sensors, LiDAR and camera.

Implementation. We will present a robust autonomous setup with a Pioneer 3-DX platform [3]. The Pioneer 3-DX is intended for research and education purposes. We use a setup as depicted in Fig. 2, left image. The robot uses rotary encoders to track the angular position and velocity of the wheels for reliable odometry data. A sonar system is integrated on the front of the chassis. The platform's main advantage is the compatibility with ROS (Robot Operating System [4]) which includes numerous resources for various tasks and applications. The complete setup is described in [5].

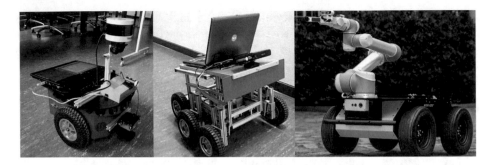

Fig. 2. Three different setups for autonomous robots. To the left there is a Pioneer 3-DX robot with a 2D-SICK LiDAR on the bottom, a 3D Velodyne LiDAR on the top and a small webcam which points to the front. In addition, the Pioneer is equipped with an array of ultrasonic sensors. In the middle there is an Fraunhofer Volksbot XL which uses a TOF camera as only sensor. The right image shows our Clearpath Husky robot, equipped with a UR5 robot arm and two Intel Realsense cameras, one on the arm, the other one in front of the platform.

The goal is to explore the environment and create a map inside an office environment. This is called SLAM: Simultaneous Localization and Mapping. The 3D-LiDAR on top of the robot is not required for the 2D mapping task. The 2D-LiDAR and adequate wheel encoders are sufficient. If ROS is used, there are different SLAM algorithms to choose from. A small selection is listed below:

GMapping [6] uses odometry data (for example wheel encoders) and a 2D-LiDAR which has to be aligned parallel to the ground. If a 3D-LiDAR has to be used, a common approach is to select a horizontal slice of its point cloud projected to a plane as input to GMapping. In the proposed setup, a SICK TiM310 LiDAR is used. The view angle is $270°$ and it has a maximum range of 4 m. The tested use case was an office building application with a long hallway with little landmarks.

Hector SLAM [7] needs only LiDAR data but can include odometry data into the algorithm. An IMU may be used to stabilize movement of the LiDAR and to filter scan points too far away from the horizontal plane.

Google Cartographer [8] uses 2D or 3D LiDAR data for SLAM. If a 3D LiDAR is used, also IMU data is required. The advancement of Cartographer over the previously mentioned algorithm is its ability to perform loop closure. This is the ability to correct drift errors in the created map after a previously visited location is visited again. Cartographer uses the concept of submaps to detect previously visited locations.

Comparison of SLAM algorithms		
GMapping	**Hector SLAM**	**Cartographer**
2D-LiDAR Odometry	2D-LiDAR opt. Odometry opt. IMU	2D-LiDAR or 3D-LiDAR opt. IMU
works well in flat office-like environments	does not depend on odometry, can account for roll and pitch of robot in uneven terrain	recognizes loop-closure, good also for outdoor usage
http://wiki.ros.org/gmapping	http://wiki.ros.org/hector_slam	http://wiki.ros.org/cartographer

Fig. 3. Properties of three SLAM algorithms which are valid options for a basic setup of an autonomous mobile robot. The literature shows many comparisons between the various LiDAR-based SLAM methods. The method should be selected based on available hardware and application.

Figure 3 shows a comparison of the three methods. A selection should be based on the available hardware setup and application environment.

If a street or track is given and the vehicle drives along the track (see Fig. 1 top row 2nd, 3rd and last image), a camera is used to identify the track. A geometry based controller (pure pursuit) is used to keep the vehicle inside the track, independently from the velocity of the vehicle, see e.g. [9,10].

If no ground odometry is available in form of wheel encoders, visual odometry is an option. A monocular visual odometry has been applied on the model ship of 1:87 scale which is depicted in Fig. 1 bottom row to the right. A full description and evaluation of the method can be found in [11]. Positioning of the ship inside a known harbour can be done using a particle filter as described here [12].

3   Creating Maps of the Environment

Task. Create a detailed map of a University Campus which can then be used for global and local navigation.

Discussion. This chapter focuses on the robot Husky A200 with a UR5 arm as depicted in the right image of Fig. 2. The robot uses a skid-steered drive which implies difficulties in measuring the odometry because of varying friction on different soil conditions. To navigate the robot, i.e. locate the robot inside a map and do path planning, a precise map is required. The robot is intended for indoor and outdoor usage. The particular use case considered is that of a bridge which can be passed on top or beneath, see Fig. 4, right image. The idea for the implementation is to use a topological map (a graph) for global planning. For local navigation, the respective local 2D map which is extracted from a global 3D map is used.

Fig. 4. The left image shows a part of the 3D map created with the described method. There is a bridge which can be used to transit from one building to another. The right image shows a photo of the same bridge at the campus of University of Applied Sciences. The bridge can be passed on top or below. Separate local 2D maps are required for local navigation of the robot, depending on its position on top or below the bridge.

The first step is therefore to create a global 3D map of the whole University Campus with sufficient resolution. This chapter describes this first step.

Implementation. It is important to mention that sensors of good quality have to be used to receive satisfying results. In this setup a Robosense RS-16 LiDAR was used, together with a Xsens IMU MTi-30. There are some algorithms available for the creation of maps.

LIO-SAM [13] was used in this project. A handheld testrig as depicted in Fig. 5 right image was constructed for mapping the campus. This allows more freedom while scanning compared to scanning with sensors mounted on the Husky A200. Crucial for good results is a close positioning of LiDAR and IMU to avoid relative motion between the two sensors. To reduce accumulating drift, LIO-SAM supports loop-closure constraints. Therefore, a path through the campus was chosen such as to create many possibilities for such constrains. Many crossings and close sections in the path can be seen on the left image in Fig. 5. The result is a 3D point cloud which represents the University Campus, see Fig. 4, left image.

Results. With decent hardware and a carefully planned path, a good quality detailed map can be created. Problems can occur close to large surfaces of glass or mirrors.

4   Creating Objects from Point Clouds

Task. The autonomous robot shall try to recognize an arbitrary small object. If recognition fails, the object shall be scanned using the depth camera, and a 3D model of the object is to be created. The goal for the robot is to successively create knowledge of objects in its environment (cups, staplers, parcels, etc.).

Implementation. The implementation is described in [14]: An Intel RealSense depth camera is used to perform a recognition and create a scan. The camera is mounted on top of the UR5 arm (see Fig. 11a or 13a).

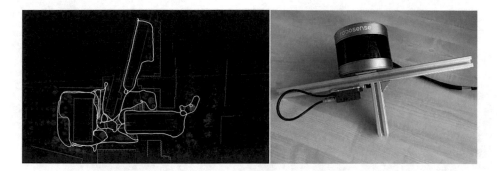

Fig. 5. The left image shows the path taken with the testrig to create the map. Frequent path crossings were included in the path to support the algorithm and improve the construction of the map with loop closures. The right image shows the testrig where the IMU is located very close to the LiDAR to avoid movement between the sensors.

The procedure starts with the object recognition part. The robot arm with the camera moves around an object in a half-circular motion and makes three 2D photos from different perspectives. All three photos are passed to a previously trained object detector. The object detector is trained with images from the *Imagenet* dataset and can initially recognize objects from 10 different classes. If all three photos detect the object to be of the same class, the object is considered to be known. If in at least one photo a different class is recognized, a subsequent scan of the object will be performed.

The scanning procedure consists of a half-circular motion of the robot arm around the object to be scanned. The depth camera creates point clouds from different perspectives of the object. The scan takes an average of 40 to 50 s.

To compose the different point clouds to a single point cloud representing the object, a registration of the point clouds has to be performed. The points in different point clouds usually do not have exact corresponding points, since the 3D points are taken from different scans. To align the point clouds with each other, approximate corresponding points need to be found and a subsequent transformation (translation and rotation) of one point cloud needs to be performed to match with the other point cloud. The algorithm used to do this is called Iterative Closes Point (ICP). See [15] for the original paper and e.g. [16] for a comparison of the different variants of ICP.

In this project, a pairwise Point-to-Plane ICP Algorithm with Pose Graph optimization is used for registration of point clouds. Corresponding points are found by finding a surface in one of the point clouds, creating a tangent plane, and calculating a normal vector, see Fig. 6a. This creates many correspondences of points which result in many different potential transformations (translation and rotation). Not all of these resulting transformations are useful and therefore a pose graph estimation is used for optimization.

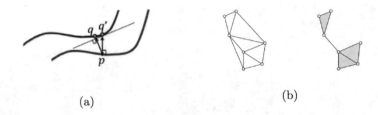

(a)   (b)

Fig. 6. (a) Visualization of point-to-plane registration (b) Example of creating α complexes

The optimization on the pose graph is performed by Levenberg-Marquardt optimization algorithm, see e.g. [17].

After the registration of collected point clouds is finished, the 3D object is created. Two different algorithms are used to create 3D meshes: Alpha Shapes (see e.g. [18]) and Poisson Surface Reconstruction (see e.g. [19]).

The Alpha Shape method requires simplices to be specified, taking into account a user defined parameter α which specifies the radius in which to look for neighbouring points. These simplices are then triangulated with Delaunay Triangulation. An alpha complex is then created. All alpha complexes created represent the desired 3D mesh.

In Fig. 6b an example of an α-complex is shown, on the left are the Delaunay triangulation sets and on the right are the complexes resulting from this triangulation.

Fig. 7. Alpha Shape 3D model creation with $\alpha = 0.005$

The Fig. 7 shows a successful alpha shape reconstruction of a Santa Claus with headphones (original: Fig. 8b). It can be seen that this method creates many defects in the model. To remove them, extra optimizations must be applied to the Polygon.

Poisson Surface Reconstruction uses an implicit function. Based on the results of a 3D indicator function $\mathbb{X} : \mathbb{R}^3 \rightarrow \{0, 1\}$, two values are distinguished: Value 1 for points within the 3D object area and 0 for the Points outside the 3D object. The reconstructed object is obtained by extracting a suitable isosurface.

The main problem of this approach is to find an appropriate function $\mathbb{X}$ whose gradient best approximates the vector field of points of the point cloud.

Figure 8a is an example of a 3D Model created with the Poison Surface Reconstruction method.

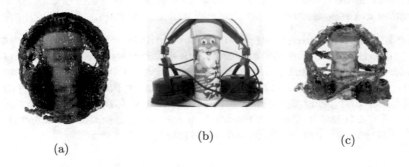

(a)   (b)   (c)

Fig. 8. (a) Object created with Poisson Surface Reconstruction (b) Original object of the Santa Claus with headphones (c) Registered 3D-Object for Santa Claus point cloud

Results. Tests were performed for six different objects which initially failed the object recognition test and were therefore to be scanned. For the majority of these objects a successful 3D model could be constructed using the described algorithms.

Figure 8a shows a successfully registered point cloud of the Santa Claus example.

37 point clouds were produced by the depth camera and were registered. Registering these point clouds with the described optimization takes approximately 150 s. After the registration, the point cloud was converted to a solid mesh using Poisson Surface Reconstruction method.

Experiments show that point cloud registration is dependent on scan speed. The speed must be chosen in such a way that the coordinates of points in the collected point clouds are sufficiently close to each other. Speed should also not be too small because then to many point clouds will be generated and all of them will be used for registration. Registration will not be significantly better, but processing time will increase.

The method does not work well with objects which are axisymmetric about the scan axis. Also transparent objects create difficulties.

5 Behaviour Inside Buildings

This chapter covers behaviour of the autonomous mobile robot inside buildings, in particular, interaction with elevators and doors. The focus is on technological aspects of recognition of elevators and doors, as well as interaction with elevator buttons and doors. Recognition of doors, interaction with doors, and the recognition of door handles are described in [20,21] and [22].

5.1 Using Elevators

Task. The autonomous mobile robot shall be able to call an elevator, press the respective buttons, enter the elevator and exit at the required floor.

Implementation. Interactions of the robot with the environment and participants are complex and require an analysis of different use cases which have been considered in [20]. Figure 9 shows maps and experiments performed in a simulation. Various difficulties arise when passing narrow doors, navigating around pedestrians, and using the UR5 arm in front of the LiDAR. An essential base ability of the robot in these scenarios is a robust recognition of elevator signs and buttons which will be discussed subsequently.

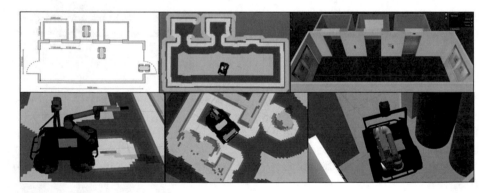

Fig. 9. Top row: Left shows a 2D model of the elevator lobby. Middle: The same model in the simulation gazebo with a costmap. The costmap is used to safely navigate the robot without hitting walls. Right: A 3D model of the elevator lobby. Bottom row: Left: The UR5 arm is in the way of the LiDAR. A costmap is created in front of the robot which prevents the robot from moving. LiDAR based navigation can therefore only be done with a folded arm, or a different LiDAR position on the robot. Middle: The costmap makes it almost impossible to drive within the narrow elevator doors or closely to people as in the right image.

Figure 10a shows different information for elevator usage to be recognized. Section 5.1 will concentrate on the recognition of elevator buttons and Sect. 5.1 will describe the procedure to press the buttons.

Recognition of Buttons. An artificial neural network is trained to recognize the elevator buttons. There are different principles of image recognition. *Classification* looks at the whole image, assuming, there is only one object in the image. *Object Detection* recognizes multiple objects inside an image and classifies these objects. This requires a two stage process: 1. Recognizing the position (in form of a rectangle) of a potential object inside the image and 2. Classifying the object inside the rectangle. There are algorithms that perform these two steps in two stages. One of the more reliable current two-stage algorithms for object detection is Faster RCNN [23]. Lately, the two stages have been combined into a single stage which led to the algorithms YOLO (You Only Look Once) [24] and SSD (Single Shot Multi-Box Detector) [25]. Both algorithms have been improved over the years. SSD has a slightly different architecture than YOLO. SSD runs well with tensorflow and tflite, which is a requirement to be able to use a hardware accelerator TPU (Tensor Processing Unit) [26]. Latest releases of YOLO can also easily be transferred to tflite for usage with TPUs. Both algorithms YOLO and SSD are actually suited quite well for inference on embedded devices like Raspberry Pi. This project uses the SSD architecture for object detection of elevator buttons.

A common way to train an object detector is to use a pre-trained SSD. The so-called Backbone component of an SSD is often pre-trained with the Microsoft COCO (common objects in context) dataset [27] or with ImageNet [28]. The bottom layers of the SSD are then trained with the application related images, elevator buttons in this case. To improve the recognition accuracy, the training images have been augmented using different lighting and viewing angles.

(a)   (b)

Fig. 10. (a) shows elevator related symbols which have to be recognized reliably. Left: Buttons to be pressed inside the elevator. Second: Position information of elevator. Third: Up-/Down-buttons on the wall in the elevator lobby. Right: Number of floor. (b) shows the result of the object detector showing the position and values of the various buttons.

Results. Recognition of the various signs and elevator buttons can be achieved quite reliably with modern artificial neural network architectures like YOLO and SSD which work well on embedded devices. Data augmentation is a crucial part of the training process in order to receive reliable results. A demonstration of a result is shown in Fig. 10b.

Pressing Buttons with Robot Arm. The UR5 robot arm has been programmed to press the respective elevator buttons. Since the camera on top of the UR5 arm is used for button recognition, basic geometry can be applied to calculate the position and movement of the arm and press the button. To apply the correct pressure of the arm while pressing the button, the integrated torque sensors of the UR5 arm may be used to define a pressure limit.

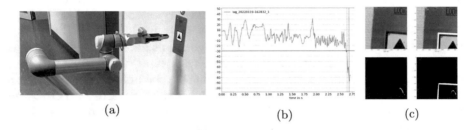

(a)   (b)   (c)

Fig. 11. (a) shows the UR5 robot arm pressing the button. (b) shows the torque measurements over time. 40 ms after the button is touched, the force exceeds the emergency stop level. (c) shows the elevator buttons with and without light and the respective recognition of the light using a filter.

Figure 11a shows the robot while pressing the elevator button. Figure 11b shows the torque over time measured within the UR5 while the arm moves towards the button. The horizontal line shows the limit for emergency stop. 40ms after the torque limit for emergency stop has been exceeded, a maximal torque of 90N has been applied and the arm is stopped automatically by the UR5 software. Experiments show, that the torque sensor in the UR5 is unsuitable for stopping at a correct pressing force. The approach of using torque sensors for stopping the arm in time had to be rejected.

A force sensor with greater sensibility at the tip of the actuator could have been used but wasn't available.

Instead, an image based method is used to stop the UR5 arm in time. The light of the button is lit when the button is pressed. The camera mounted on top of the arm recognizes the light of the button as shown in the images in Fig. 11c. Basic image recognition methods from OpenCV [29] may be used for this task.

Results. Pressing buttons with the UR5 arm remains a difficult task. The proposed solution is still unreliable. The torque sensors inside the arm may not be used for the discussed scenario. Some elevator buttons in the building have broken lights. The proposed method clearly doesn't work in this case.

5.2   Recognizing Doors

Task. A robust recognition of doors in various states (open, half-open, closed) is to be developed, based on object detection. Suitable data augmentation of the training data shall be applied to achieve a robust recognition. Various methods for data augmentation are to be compared.

Implementation. In [22] 110 images of doors were taken at the University of Applied Sciences Hamburg in an office-like environment. Several data augmentation methods were applied:

1. For the brightness dataset the degree of brightness was adjusted from low to high using the python package Augmentor [30]. See Fig. 12a.
2. The same procedure was applied for color. See Fig. 12b.
3. The same procedure was applied for contrast. See Fig. 12c.
4. Parts of the images were erased/deleted using a range from few to many cut outs. See Fig. 12d.
5. A new method called greenscreen was developed in this work. For images containing open and semi-open doors the area in the frame was cut out and replaced with green pixels. The resulting greenscreen was replaced with noise images. See Fig. 12e.

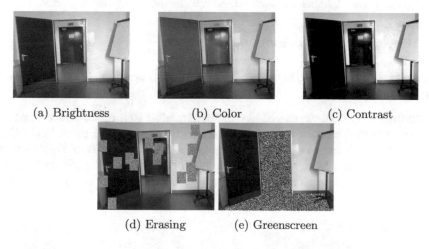

(a) Brightness   (b) Color   (c) Contrast

(d) Erasing   (e) Greenscreen

Fig. 12. Example images after applying image data augmentation methods

Different datasets were created using only one augmentation method respectively. An artificial neural network was trained with each of the datasets and the results were compared using different metric values.

Results. Data augmentation by increasing the contrast of images leads to the highest positive distance in comparison of metric values in contrast to the default image dataset. Erasing image details, changing color, changing brightness (in this order) have fewer impact. The greenscreen method has the lowest impact in these experiments.

Table 1. Detection rates of the default (D), combined (C), DeepDoors2 (DD2), default+combined and default+combined+DeepDoors2 dataset at the front of a door. Magenta colored values are the highest value in the given row.

	D	C	DD2	D+C	D+C+DD2
Open	50%	66%	50%	33%	33%
Semi	50%	83%	50%	66%	83%
Closed	83%	100%	100%	100%	100%
Total	61%	83%	66%	66%	72%

A new combined (C) dataset containing all augmented images was created. Utilizing the combined dataset of all previously mentioned image data augmentation methods leads to the highest detection rate in comparison to the other training datasets, see Table 1 column C. An external dataset DeepDoors2 [31] (DD2) which was used as a benchmark was outperformed by the combined (C) dataset. Adding DD2 to the combined dataset C (D+C+DD2) didn't improve the detection rate further but showed worse performance on open doors.

5.3   Recognizing Door Handles

Task. Door handles shall be recognized reliably to provide a basis for gripping with the robot arm.

Implementation. A stereo vision camera mounted to a robot arm is used. It is rotated vertically, because this causes the handle to stand out more in the depth image. The RGB image of the stereo camera is processed by a YoloV5 neural network, which has been trained to detect door handles. The object detection provides a rectangle which is used as a region of interest (ROI) in which the

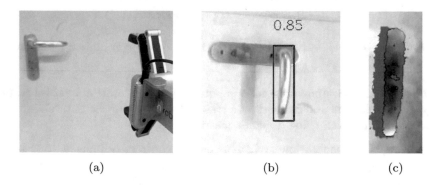

(a)   (b)   (c)

Fig. 13. In (a) a third person view of the gripper is shown. The realsense camera is pointing towards the handle. (b) shows the result of the handle detection. In (c) the ROI of the depth image is displayed.

handle is assumed. The ROI is then applied to the depth image, from which the handle surface can be identified.

Results. Quantitative results have not yet been obtained. The investigations serve as a technological feasibility study. YoloV5 has proven to be useful for these applications.

5.4   Opening Doors

Task. The robot shall use its arm to open doors. An experimental setup is to be constructed which is suitable for further experiments.

Implementation. In [21] a method is described to open doors with an autonomous mobile robot using a UR5 robot arm. The method is implemented and tested in a ROS gazebo simulation.

The door opening process has been analysed and broken down into 3 steps, which have several substeps.

1. Positioning of the robot in front of the door, detection of the door, detection of the door handle
2. Gripping the door handle with the robot arm and unlock the door
3. Opening the door by pulling

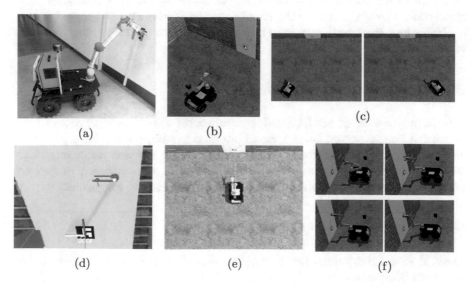

Fig. 14. Opening doors in a simulation. (a) shows the real robot positioned in front of a door. (b) and (c) shows the robot in the simulation recognizing an ArUco marker, which helps positioning the robot without having to recognize the actual door. (d) shows the recognition process of the door handle in the simulation. After detection of the door handle, the robot needs to be repositioned to the side of the door which holds the door handle (e). (f) shows the pulling process.

Only step 1 shall be discussed here. Steps 2 and 3 apply predefined processes and are of less interest.

The process has been modeled using *behaviour trees*, see e.g. [32]. The process begins with the detection of an ArUco marker which has been attached to the door (for ArUco Markers see e.g. [33]), see Fig. 14b. ArUco Markers can easily be detected in a camera image using OpenCV [29]. The data from the depth camera is then used to calculate the exact position of the door and the orientation of the robot to the door using basic geometry. The goal position of the robot in front of the door is calculated and move_base is used to drive to this position (see Fig. 14c). Using depth information of the camera, the opening orientation of the door is evaluated. Afterwards the point cloud of the realsense is used to detect the door handle. The corresponding coordinates for the positioning of the robot arm are calculated. Figure 14d shows the coordinate system in the center of the door and the rotational axis of the handle. The detection is based on the detection of jumps and edges in the point cloud. After detection of the door handle the robot needs to be repositioned on the side of the door with the handle, see Fig. 14e.

Results. A fundamental problem with this approach are the difficulties of positioning the robot in front of the door precisely. Due to inaccuracies in odometry, which are unavoidable in skid-steering, and due to the uniform environment, the position and orientation of the robot cannot be determined precisely and therefore exact positioning is difficult. The method works occasionally in the simulation. However, even in the simulation the method is not very reliable.

Behaviour trees have proven to be a good utility to model complex behaviour of the robot.

6   Man-Robot Interaction

Several projects have been initiated for interaction of humans with robots, two of which shall be presented here: A visual interaction component with a pedestrian, and an acoustical interaction component for emergency voice commands. Other ongoing projects which are not described here are a teleoperation project over 5G mobile network and an augmented reality project to display sensory data.

6.1   Visual Interaction Component

Task. A visual component for the mobile robot shall be constructed which interacts with pedestrians.

Implementation. An LED light strip was used to react to persons which are detected via LiDAR. Some requirements were defined for the interaction system.

1. The robot indicates that it detects people in its environment.
2. The robot displays its internal system status.
3. The robot indicates that it wants to perform a turn.

An LED strip of the type WS2812B with individually addressable LED modules was used. This allows each module to shine in a different color. The LiDAR used for person detection was a RS-Lidar-16. Using the results of [34], persons were detected in a LiDAR point cloud. The detected persons are extracted by the interaction system and assigned to segments on the LED strip. These light up as shown in Fig. 15.

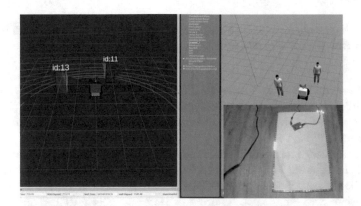

Fig. 15. Demonstration of person recognition with the visualization on the LED strip. On the left side the detection of the persons on the LiDAR by [34] can be seen. On the bottom right, the detected persons are shown on the LED strip [35].

Results. An available software for detecting persons in LiDAR point clouds is used to enable the mobile robot to react in a visual way. The project was tested in a simulation and in an experimental setup and works as expected.

6.2 Simple Voice Commands

Task In the voice control project [36] the robot shall respond to simple one-word voice commands. The basic idea is to have a voice recognition which is independent from a connection to the internet and can recognize emergency stop commands. The software shall support an easy way for localization (change to another language).

Implementation. An artificial neural network has been trained to recognize a set of basic commands. The commands have been chosen such that an arbitrary person could intuitively trigger an emergency stop of the robot (Stop, Halt, Turn Off, etc.).

The hardware components of the project are shown in Fig. 16. A google Coral Tensor Processing Unit (TPU) [26] is used to accelerate the inference. The extension board is used as a an USB hub to connect the devices with the RPi Zero, and to connect the RPi Zero with the robot control computer via Ethernet.

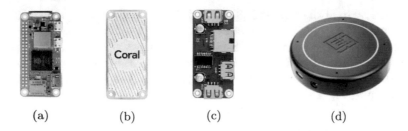

(a)   (b)   (c)   (d)

Fig. 16. Components for voice control. Raspberry Pi Zero, Google Coral TPU, Waveshare USB Hub Hat, ReSpeaker Mic Array.

As an artificial neural network architecture a DS-CNN (Depthwise Separable Convolutional Neural Network) as suggested in [37] has been used. This architecture proves to be a good compromise between accuracy and suitability for edge devices for *keyword spotting*.

The dataset for training the artificial neural network needs to contain speech commands in different languages. The Google Speech Command Dataset [42] and other available speech command datasets mostly contain data only in English or in another few languages. Therefore Text-to-Speech (TTS) methods were applied for training. TTS creates synthetic speech in almost all languages and allows automation of the training process. This saves a lot of time training the artificial neural network in various languages. Details on TTS providers are given in [36].

Results. Training artificial neural networks with TTS for localization allows automation of the training process. However, the robustness of the recognition when TTS was used for training is not satisfying. Arbitrary human speakers were not recognized well. This is probably due to the synthetic character of the training data. Good results were obtained in the English language when the Google Speech Command Dataset was used for training.

7   Positioning with Reinforcement Learning

Task. A positioning system for usage inside buildings and outside between high buildings is to be developed. Precision of the positioning system should be as good as possible, preferably much below 50 cm.

Implementation. In [38] an ultra-wideband wireless technology (UWB) is used for positioning of autonomous mobile robots. Three methods for improvement of precision have been implemented and compared.

UWB describes electromagnetic signals with an absolute bandwidth of at least 500 MHz or a fractional bandwidth of higher than 20%. Signals are usually transmitted in very short pulses with a low duty cycle. The transmitted power is distributed over the entire frequency range so that the signal shape resembles the background noise. In many regions, the broad spectrum from 3.1 to 10.6 GHz is available for unlicensed use up to a maximum power density of -41.3 dBm/MHz. These characteristics, along with the high bandwidth and frequencies, enable very accurate and high temporal resolution distance measurements in short-range wireless networks. In addition, the cost and power consumption of a UWB transceiver are low.

A hardware (see Fig. 17, left image) and software platform for an in-house UWB localization system for customized use in University projects and research groups is developed.

To locate a node in such a wireless network, signals are exchanged between a number of stationary reference nodes. This involves determining the distance to each reference node by measuring the time of flight between the transmitter and receiver of the signal. The position of the tracked object can then be localized by trilateration at the intersection of the circles around the reference nodes with the radii of the corresponding distances (see Fig. 17, right image).

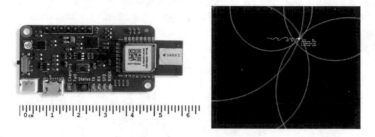

Fig. 17. The left image shows the PCB of our designed UWB beacon based on Qorvo's DWM1000 UWB transceiver. The right figure shows a screenshot of the central localization unit for trilateration calculation. The distance circles intersect at roughly the same point where the tracked target is located.

From GPS applications, Kalman filter or particle filter based techniques are known to improve the reliability and accuracy of position determination. For the UWB positioning system, a novel approach based on [39] to improve localization accuracy is examined. This approach uses reinforcement learning (RL) to learn an optimal policy for correcting position measurements. See [40] for an introduction to Reinforcement Learning.

An action of the RL agent is defined as an operation to adjust x- and y-coordinates of the measured position. To reduce complexity, corrective actions are discretized into 121 small steps. Thus, there are 5 possible steps in each positive and negative x- and y-direction by which the position can be shifted. The model predicts a corrective action based on a fixed size observation vector containing the current measured position as well as a set of historic position estimations. The observation vector is updated continuously in a sliding window approach. When the UWB localization system reports a new position, it is appended to the vector, replacing the oldest position estimate. This modeling of the observation space allows the model to further incorporate information about the past trajectory.

Of particular interest from a reinforcement learning perspective is the mechanism employed to calculate the reward. One challenge is that the ground truth position is unknown and therefore cannot be used to form the reward function. To solve this problem, an abstract reward mechanism based on a confidence measure of multiple model predictions is used. When a new position is measured, the model is trained k times rather than just once. So based on the current model parameters, k possible corrective actions are predicted. A confidence ellipse is formed using the k predictions, whose size provides an indication of the certainty of the RL agent's predictions. This confidence measure is used instead to build the reward function.

The Asynchronous Advantage Actor Critic (A3C) [41] architecture is used to train the RL model. A3C is a policy-gradient algorithm that learns both a strategy and a value function. In an asynchronous training architecture, multiple agents run in parallel in different threads. Each agent maintains its own independent copy of the training environment and trains a separate model. The advantages of A3C are shorter training times with more stable results compared to more traditional RL algorithms.

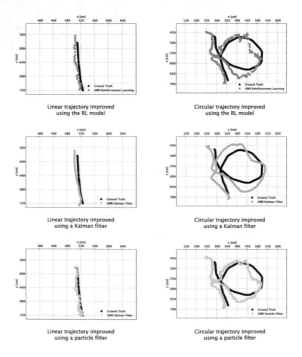

Fig. 18. Comparison of the presented RL model, a simple Kalman and particle filter to correct two trajectories sampled by our UWB localization system. The ground truth is drawn in black.

Results. To benchmark the performance of the presented RL correction method, the accuracy is compared with a simple Kalman and particle filter. Figure 18 shows the results. A camera-based AR tracking system provides a ground truth to the examined trajectories.

The experimental validation reveals that reinforcement learning can be utilized to improve the localization accuracy besides established filter and correction algorithms. The proposed RL model shows comparable performance to the particle filter and is particularly suitable for non-linear motions. In real world applications, the presented UWB localization system achieves an accuracy of up to 10 cm, making it appropriate for use in typical application scenarios.

8   Final Remarks

All projects were carried out and implemented by individual researchers and students as part of a working group of around 20 software, sensor and AI specialists. The great mutual support on various problems with hardware or software has contributed significantly to the success of the individual projects.

Ongoing efforts are towards integrating the various capabilities of the robot into a joint software structure. The robot shall be able to plan global tasks, divide them into subtasks, and resolve difficulties which arise on the way.

References

1. Pareigis, S., Tiedemann, T., Stelldinger, P., Schönherr, N., Mang, M.: autosys research lab (2022). https://autosys.informatik.haw-hamburg.de. Accessed 30 Nov 2022
2. Tiedemann, T., Schwalb, L., Kasten, M., Grotkasten, R., Pareigis, S.: Miniature autonomy as means to find new approaches in reliable autonomous driving AI method design. Front. Neurorobot. **16**, 846355 (2022)
3. Adept MobileRobots: Pioneer and Pioneer-compatible platforms (2022). http://wiki.ros.org/Robots/AMR_Pioneer_Compatible. Accessed 30 Nov 2022
4. ROS: Robot Operating System (2022). https://www.ros.org. Accessed 30 Nov 2022
5. Casagrande, M.: Robust setup of an autonomous mobile robot research platform with multi-sensor integration in ROS. B.S. thesis, University of Applied Sciences Hamburg (2018)
6. ROS: gmapping (2022). http://wiki.ros.org/gmapping. Accessed 30 Nov 2022
7. ROS: hector SLAM (2022). http://wiki.ros.org/hector_slam. Accessed 30 Nov 2022
8. ROS: cartographer (2022). http://wiki.ros.org/cartographer. Accessed 30 Nov 2022
9. Jenning, E.: Systemidentifikation eines autonomen Fahrzeugs mit einer robusten, kamerabasierten Fahrspurerkennung in Echtzeit. B.S. thesis, University of Applied Sciences Hamburg (2008)
10. Nikolov, I.: Verfahren zur Fahrbahnverfolgung eines autonomen Fahrzeugs mittels Pure Pursuit und Follow-the-carrot. B.S. thesis, University of Applied Sciences Hamburg (2009)
11. Stark, F.: Monokulare visuelle Odometrie auf einem autonomen Miniaturschiff. B.S. thesis, University of Applied Sciences Hamburg (2020)

12. Schnirpel, T.: Lokalisierung eines Miniaturschiffs durch Distanzsensoren. University of Applied Sciences Hamburg, Internal paper (2020)
13. Shan, T., Englot, B., Meyers, D., Wang, W., Ratti, C., Daniela, R.: LIO-SAM: tightly-coupled lidar inertial odometry via smoothing and mapping. In: IEEE/RSJ International Conference on Intelligent Robots and Systems (IROS), pp. 5135–5142 (2020)
14. Mihajlov, D.: Autonome 3D-Objektmodellierung durch aktive Szenenexploration mit einem UR5 Greifarm. B.S. thesis, University of Applied Sciences Hamburg (2021)
15. Besl, P.J., McKay, N.D.: A method for registration of 3-D shapes. IEEE Trans. Pattern Anal. Mach. Intell. **14**, 239–256 (1992)
16. Wang, L., Sun, X.: Comparisons of iterative closest point algorithms. In: Park, J.J.J.H., Pan, Y., Chao, H.-C., Yi, G. (eds.) Ubiquitous Computing Application and Wireless Sensor. LNEE, vol. 331, pp. 649–655. Springer, Dordrecht (2015). https://doi.org/10.1007/978-94-017-9618-7_68
17. Darcis, M., Swinkels, W., Güzel, A.E., Claesen, L.: PoseLab: a levenberg-marquardt based prototyping environment for camera pose estimation. In: 2018 11th International Congress on Image and Signal Processing, BioMedical Engineering and Informatics (CISP-BMEI), pp. 1–6 (2018). https://doi.org/10.1109/CISP-BMEI.2018.8633112
18. Xu, X., Harada, K.: Automatic surface reconstruction with alpha-shape method. Vis. Comput. **19**(7), 431–443 (2003). https://doi.org/10.1007/s00371-003-0207-1
19. Kazhdan, M.M., Bolitho, M., Hoppe, H.: Poisson surface reconstruction. In: Eurographics Symposium on Geometry Processing (2006)
20. Denecke, E.: Autonome Nutzung von Aufzügen durch einen Industrieroboter. B.S. thesis, University of Applied Sciences Hamburg (2022)
21. Kasper, J.: Autonome Erkennung und Interaktion mit Türen in der Robotik. B.S. thesis, University of Applied Sciences Hamburg (2022)
22. Tran, J.: Robust door detection using deep learning in visual based robot navigation. B.S. thesis, University of Applied Sciences Hamburg (2022)
23. Ren, S., He, K., Girshick, R.B., Sun, J.: Faster R-CNN: towards real-time object detection with region proposal networks. In: NIPS, pp. 91–99 (2015). http://dblp.uni-trier.de/db/conf/nips/nips2015.html
24. Redmon, J., Divvala, S.K., Girshick, R.B., Farhadi, A.: You only look once: unified, real-time object detection. CoRR (2015). https://arxiv.org/abs/1506.02640
25. Liu, W., et al.: SSD: single shot multibox detector. CoRR (2015). https://arxiv.org/abs/1512.02325
26. Google: Coral (2022). https://coral.ai. Accessed 30 Nov 2022
27. COCO Common Objects in Context (2022). https://cocodataset.org/. Accessed 30 Nov 2022
28. ImageNet (2022). https://www.image-net.org. Accessed 30 Nov 2022
29. Bradski, G.: The openCV library. Dr. Dobb's J. Softw. Tools **25**(11), 120–123 (2000)
30. Bloice, M.D., Stocker, C., Holzinger, A.: Augmentor: an image augmentation library for machine learning. arXiv (2017). https://doi.org/10.48550/ARXIV.1708.04680. https://arxiv.org/abs/1708.04680
31. Ramôa, J.G., Lopes, V., Alexandre, L.A., Mogo, S.: Real-time 2D–3D door detection and state classification on a low-power device. SN Appl. Sci. **3**(5), 1–13 (2021). https://doi.org/10.1007/s42452-021-04588-3
32. Colledanchise, M., Ögren, P.: Behavior trees in robotics and AI: an introduction. CoRR (2017). https://arxiv.org/abs/1709.00084

33. Avola, D., Cinque, L., Foresti, G.L., Mercuri, C., Pannone, D.: A practical framework for the development of augmented reality applications by using ArUco markers. In: Proceedings of the 5th International Conference on Pattern Recognition Applications and Methods - ICPRAM, pp. 645–654 (2016). https://doi.org/10.5220/0005755806450654

34. Koide, K., Miura, J., Menegatti, E.: A portable three-dimensional LIDAR-based system for long-term and wide-area people behavior measurement. Int. J. Adv. Robot. Syst. **16** (2019). https://doi.org/10.1177/1729881419841532

35. Abdelkarim, A.: Human-machine interaction on an autonomous robot through visual signals. B.S. thesis, University of Applied Sciences Hamburg (2022)

36. Sentler, D.: Detektion von Sprachbefehlen auf Edge-Geräten unterstützt durch automatisierte Trainingsdaten-Synthese für eine Not-Halt-Anwendung. B.S. thesis, University of Applied Sciences Hamburg (2022)

37. Zhang, Y., Suda, N., Lai, L., Chandra, V.: Hello edge: keyword spotting on microcontrollers. CoRR (2017). https://arxiv.org/abs/1711.07128

38. Koch, S.: Hochgenaue Positionsbestimmung mit Ultra-wideband und Reinforcement Learning. B.S. thesis, University of Applied Sciences Hamburg (2021)

39. Zhang, E., Masoud, N.: Increasing GPS localization accuracy with reinforcement learning. IEEE Trans. Intell. Transp. Syst. **22**(5), 2615–2626 (2021)

40. Sutton, R.S., Barto, A.G.: Reinforcement Learning: An Introduction. MIT Press, Cambridge (2018)

41. Mnih, V., et al.: Asynchronous methods for deep reinforcement learning (2016). https://arxiv.org/abs/1602.01783

42. Warden, P.: Speech commands: a dataset for limited-vocabulary speech recognition (2018)

Neural Networks in View of Explainable Artificial Intelligence

Wolfgang A. Halang[1]([✉]), Maytiyanin Komkhao[2], and Sunantha Sodsee[3]

[1] FernUniversiät in Hagen, 58084 Hagen, Germany
wolfgang.halang@fernuni-hagen.de
[2] Rajamangala University of Technology Phra Nakhon, 10800 Bangkok, Thailand
[3] King Mongkut's University of Technology North Bangkok, 10800 Bangkok, Thailand

Abstract. Artificial neural networks are just mathematical functions used for approximation purposes, and learning is a euphemism for determining the parameters of these functions. Experience from other areas of approximation theory shows that closed-form approximation functions determined globally such as neural networks tend to be quite rippled and are, thus, outperformed by approximation functions defined locally. With respect to artificial intelligence, neural networks lack explainability. Therefore, research in novel approaches for local approximators apt to replace neural networks and to outperform them with respect to accuracy, computational expense and explainability is suggested.

Keywords: Neural networks · universal, global and local approximation · artificial intelligence · explainability

1 Introduction

It must have been in the late 1980s or early 1990s that I heard more and more often the term neural networks then unknown to me. Eventually, I had enough and looked it up. I had to laugh, because I knew almost everything on neural networks. In the summer semester 1976 I had sat, namely, at the (now Technical) University of Dortmund with only three fellow students in a special course. This course's name was rather unsexy explaining the low number of attendees: *threshold logic* [3]. My interest in hardware, which is not very pronounced in computer science, had induced me to take the course, but its content was essentially plain mathematics.

The fundamental elements of threshold logic are threshold gates. Such a gate g compares a linear combination of real-valued inputs with a threshold s and yields as output the binary information whether the value of the linear combination exceeds the threshold:

$$g \colon \begin{cases} \mathbb{R}^n & \to \{0,1\} \\ (x_1,\ldots,x_n) & \to \sum_{i=1}^n a_i \cdot x_i + b \geq s \ , \ a_i, b \in \mathbb{R}, \ i = 1,\ldots,n \end{cases}$$

H. Unger and M. Schaible (Eds.): Real-Time 2022, LNNS 674, pp. 146–150, 2023.
https://doi.org/10.1007/978-3-031-32700-1_15

In principle, this is all: a class of rather simple mathematical functions. As long as they were still called threshold gates they just evoked boredom. This changed abruptly by simple renaming: threshold gates turned to neurons or, particularly overblown, even to perceptrons. The threshold gates were a little bit generalised by means of replacing the threshold comparisons downstream the linear combinations by real-valued functions called activation functions. Interconnections of several such neurons were finally called (artificial) neural networks, in order to suggest analogies with the human brain, its functionality, and with intelligence as well as to fuel a hype. Since then, neural networks are employed for approximation purposes. Consistent with a nomenclature, which shall evoke conceptions of cognitive processes, the determination of approximating neural networks' coefficients is called learning.

2 Neural Networks as Universal Approximators

For centuries it is known that any continuous, real-valued function on a compact subset of $\mathbb{R}^n$, $n \geq 1$, can arbitrarily well be approximated by step functions. For this, the domain is partitioned into small regions, and for each of them a step function's value is set as a constant determined locally. That is the common practice of sampling, discretisation, quantisation and digitalisation. Such piecewise constant approximation functions can also be represented as linear combinations of threshold gates of type g as defined above [5].

In general, of functions to be approximated only their values for finite numbers of arguments are known. Hence, an approximation problem consists in finding the element of a given class of functions whose values at mutually different locations v_i, $i = 1, ..., k$, $k \geq 1$, in n-dimensional space come as close as possible there to given values w_i, $i = 1, ..., k$, with respect to a selected distance metric. Often, the data w_i are measured values corrupted by noise. To keep the influence of noise small, the elements of a function class employed for approximation must, therefore, be continuous. Utilising neural networks, this requires activation functions to be continuous. They must even be differentiable, if the coefficients occurring in the interconnections of neurons are to be determined by efficient iteration algorithms, such as the gradient method. For this reason, usually sigmoid functions are chosen as activation functions in neural networks, whose values approach 1 (0) for increasing positive (negative) arguments. The narrower the transition region is between their values equal or close to 0 and equal or close to 1, the more similar are sigmoid functions to the discontinuous step function realising the threshold comparison. Thus, it is by no means surprising that any continuous, real-valued function on a compact subset of $\mathbb{R}^n$, $n \geq 1$ can, under the Čebyšev norm, be approximated arbitrarily well by linear combinations of sufficiently many neurons according to the Universal Approximation Theorem of Cybenko [1]. First proven for sigmoid-like activation functions, this result was extended by Hornik to any activation functions [2].

The function values of approximating neural networks are, due to the continuity of their sigmoid or other activation functions, not determined locally anymore, but by all given data points (v_i, w_i), $i = 1, ..., k$. As the complete set of points needs to be taken into account for the calculation of an approximating neural network's parameters, the effort required for this is very high, particularly when applying iterative methods. In the end, however, it is neither known whether a best — or just a good — approximation according to the chosen distance metric was found, nor is there information on the approximation quality. Accordingly, Cybenko concludes: "The important questions that remain to be answered deal with feasibility, namely how many terms in the summation (or equivalently, how many neural nodes) are required to yield an approximation of a given quality? ... At this point, we can only say that we suspect quite strongly that the overwhelming majority of approximation problems will require astronomical numbers of terms. This feeling is based on the curse of dimensionality that plagues multidimensional approximation theory and statistics" [1].

To conclude this section, although neural networks are considered one of the major techniques of artificial intelligence (AI), "from the Universal Approximation Theorem, we understand that neural networks are not really intelligent at all, but just good estimators hidden under a guise of multidimensionality" [5].

3   What Do We Really Want?

So far we have seen that neural networks are a tool of approximation theory. It is questionable, however, that this tool is suitable for AI applications as well, because just values sampled at certain points in space, but no functions are to be approximated there. Since there is no information on the behaviour of the sampled entity between the sampling points, the property assumed and actually desired in AI applications is *similarity*: If a point is — in a certain sense — close to a member of a given data set, their values ought to be close, too. Neural networks have, however, two main shortcomings preventing to achieve this objective:

High variability. This phenomenon is well known from polynomial approximations. Since continuous approximation functions of a certain class cannot take any form, they tend to be ripply with big variations in order to meet their globally given approximation conditions. In contrast to this, local approximation methods stand out due to many favourable properties such as low ripple. To determine such approximation functions, these methods employ just relatively few data points close to the locations where function values are sought.

Non-explainability. After — mainly iterative — determination of a neural network's parameters, the relation between the original data and the calculated weights is totally blurred. This phenomenon gave rise to the desire for "explainable artificial intelligence". In contrast to globally defined neural networks, it can easily be explained and understood how a local approximation is constructed, and how the value to be associated with a point is derived from the values sampled at neighbouring points.

4   Future Work

Based on the considerations above, it is suggested to devise and evaluate multidimensional approximation operators constructed locally, with the approximation functions' parameters depending in easily explainable ways on the underlying data sets, for their ability to replace neural networks in AI applications. To this end, the computational costs of constructing neural networks ("learning") and their approximation quality should be compared empirically with the ones of a variety of local approaches on the basis of realistic data sets. Particularly investigated should be whether the big effort required to determine multilayer neural networks ("deep learning") is worthwhile at all. With respect to define novel local approximation operators one can let one's imagination run wild. To start with a few obvious ideas are mentioned below.

Let in n-dimensional space, $n \geq 1$, a point p and a set Q of further $k \geq 1$ mutually different next, neighbouring or other points in distances r_i from p with corresponding function, measuring or other values w_i, $i = 1, ..., k$, be given. Then, the approximation effort and quality of associating the following values with p is to be investigated:

– **For $k = n + 1$ evaluate at location p the hyperplane defined by the k points of Q**
  This approach requires to solve a system of $n + 1$ linear equations. For $n = 1$ it reduces to linear interpolation between two points, i.e. the method of choice in case there is no knowledge of the approximated entity between given data points.
– **Assign to p the value of the closest point in Q:**

$$w := w_j, \; r_j = Min\{r_i \mid i = 1, ..., k\}$$

This is the simplest approach, but it is not unlikely that it may already be the best: "Any continuous function on a compact set can be approximated by a piecewise function. ... Cybenko was more specific about this piecewise function, however, in that it could be constant, essentially consisting of several steps fitted to the function. With enough constant regions (steps), one can reasonably estimate the function within the bounds it is given in" [5]. Furthermore, this representation is nothing else than threshold logic, i.e. the original form of artificial neural networks.
– **Assign to p linear combinations of Q's values:**

$$w = \sum_{i=1}^{k} b_i \cdot w_i$$

with $\sum_{i=1}^{k} b_i = 1$.
This approach is suggested by the calculations performed in neurons with the identity as activation function. For instance, the coefficients b_i may be

selected as follows, expressing weighting with respect to distance between p and the members of Q:

$$b_i = \frac{1}{k-1}\left(1 - \frac{r_i}{\sum_{j=1}^{k} r_j}\right)$$

or

$$b_i = \frac{r_{k+1-i}}{\sum_{j=1}^{k} r_j}$$

for r_i, $i = 1, ..., k$, in monotonously increasing order.

References

1. Cybenko, G.: Approximation by superpositions of a sigmoidal function. Math. Control Signals Syst. **2**(4), 303–314 (1989)
2. Hornik, K.: Approximation capabilities of multilayer feedforward networks. Neural Netw. **4**(2), 251–257 (1991)
3. Hurst, S.L.: Threshold Logic. Mills & Boon Ltd, London (1971)
4. Kratsios, A.: The universal approximation property – characterization, construction, representation, and existence. Ann. Math. Artif. Intell. **89**, 435–469 (2021). https://doi.org/10.1007/s10472-020-09723-1
5. Ye, A.: You Don't Understand Neural Networks Until You Understand the Universal Approximation Theorem – The Proof Behind the Neural Network's Power. https://medium.com/analytics-vidhya/you-dont-understand-neural-networks-until-you-understand-the-universal-approximation-theorem-85b3e7677126

Reconstruct a Distributed Co-occurrence Graph in a P2P Network Without Overhead

Martin Drebinger$^{(\boxtimes)}$, Oliver Tominski, and Herwig Unger

Communication Networks, FernUniversität in Hagen, Unistr. 1, 58084 Hagen, Germany
drebinger@mailbox.org

Abstract. The distributed storage of a co-occurrence graph in a peer-to-peer (P2P) network faces the loss of knowledge when peers disappear. Recovery mechanisms are necessary to ensure stable operation, but conventional redundancy methods have too many drawbacks in the P2P network.

It was possible to show that using messages flowing through the rest of the network provided enough redundancy for comprehensive recovery of a faulty and failed peer. The messages were created in normal operation and can be used without much additional (network) load. There is a high probability that the network will continuously produce valid results, even if peers disappear and have to be replaced by new, initially empty peers. Thus, a co-occurrence graph can be stored in a peer-to-peer network in a very robust manner, and the algorithms found can significantly increase fault tolerance.

Keywords: fault tolerance · redundancy · co-occurrence · graph · peer-to-peer · network

1 Motivation

Avoidance of data loss is desirable in nearly every information system. In distributed systems – especially without an orchestrator – it can be a very expensive task to provide data safety. Distributed systems often use a variable base of underlying computers. In peer-to-peer networks, this is almost always the case. In contrast to operation of a normal server, the loss of components is not a remarkable situation. It is commonplace and must be handled by protocols. If the protocols are well-designed, a peer to peer network can be a resilient system with high computing power and scalable storage. The distributed system considered here is a peer-to-peer network, as described in [2]. The global knowledge is stored in the form of a co-occurrence graph, where words are used as nodes. The words of the graph are additionally connected with data worth knowing. The graph itself is stored on a decentralized distributed system consisting of many peers. It is inevitable that participating peers may drop out, resulting in a loss

© The Author(s), under exclusive license to Springer Nature Switzerland AG 2023
H. Unger and M. Schaible (Eds.): Real-Time 2022, LNNS 674, pp. 151–170, 2023.
https://doi.org/10.1007/978-3-031-32700-1_16

of knowledge as a direct consequence. That is why a method of fault-tolerance must be established for these events in order to minimize data loss.

Is it possible to gather enough information from the messages flowing through the co-occurrence graph to reconstruct the lost parts and resume normal operations?

One idea of nature for a quite similar problem can be found in the brain. If parts of the system fails, e.g. after a stroke, it is often possible that (at least parts) of the functionality of the failed area can be taken over by unaffected areas after a time [1]. It should be verified whether these mechanisms inspired by this recovery capability also work in the case of a peer-to-peer network presented here.

The experiments conducted with the help of a simulation shows that some problems occur. The worst one is that many nodes are reconstructed that were not part of the lost subgraph. It should be shown that there are more complex but much more powerful methods to achieve fast reconstruction without adding too many nodes. The main idea is not to use the messages individually, but to collect them to gain more information.

2   Description of the Decentralized Environment

In this chapter, the specific network in which this work was conducted is described. The network has a large inherent redundancy of stored knowledge that is harnessed by our reconstruction approaches.

2.1   Distributed System

The distributed system consists of peers $P_0 \ldots P_n$. Each peer P_x includes a local graph L_x and a part (G_x) of the (virtual) global graph G, which nowhere exists as a whole. This global graph G is a superposition of all local graphs and contains all nodes and edges from these graphs. The global subgraphs on each peer are connected by outgoing edges. A node represents a word that can exist in more than one local graph L_x, but should exist only once in all the global subgraphs G_x. The article [4] describes mechanisms to achieve this condition without a central control unit and the Fig. 1 shows this basic architecture of the network.

2.2   Routing

The routing procedure is described in more detail in [5]. Here, only the points that are important for message generation and thus for reconstruction based on messages are explained.

When a peer tries to access information associated with a node in the global graph, there are two possible situations: In the trivial case, the node N_k is located in the global subgraph G_i on the same peer P_i and the node can be easily found and the information can be retrieved. If the node is stored in the global subgraph of another peer, some facts can be used to find this node:

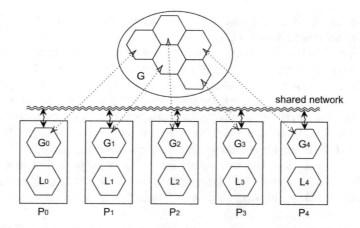

Fig. 1. Peer to peer network where each peer P_x hosts two graphs: a local graph and a part of the global graph

- a peer P_i can only search for a node N that is known to itself in its local graph ($N \in L_i$)
- every path in the local graph L_i is also contained in the global graph G
- the local graph is connected, so between any two nodes N, N' in G there exists at least one path p that connects them ($\forall(N, N')\exists p_{N,N'}$)
- some (at least one) nodes of the peer's local graph are also included in its global subgraph ($G_i \cap L_i \neq \emptyset$)

To use the knowledge stored in the global graph, the following steps are required to find any node from the local subgraph in the global graph of another peer.

1. find the selected node N_x on the peer P_i local graph L_i
2. find a (ideally short) path p_{N_x,N_y} in the local graph to a node N_y that is also in the global subgraph of the peer ($N_y \in G_i \cap L_i$)
3. create a message whose path starts at the last node N_y (the one on the global subgraph on the same peer) and contains path p_{N_x,N_y} in reverse order (message path: p_{N_y,N_x}). So the message can now follow its path in the global graph G to node N_x, which exists due to preconditions mentioned above.
4. upon reaching the last node in the message path, the associated peer sends the node's information linked to it back to the peer P_i that initiated the search.

This procedure always works under the constraints specified above. The path used does not necessarily have to be the shortest possible in the global graph.

2.3 Messages

As previously described, messages are needed for finding resources in the network. Each message contains all intermediate nodes to progress step by step between origin and destination $p_{N_0,N_x} = (N_0, N_1, \ldots, N_{x-1}, N_x)$. With the help of the imprinted path p_m, a message traverses the global graph G without noticing that it is distributed among different peers. The peers themselves only know their direct neighbors, with whom they are connected via outgoing edges, and can forward messages accordingly.

If a peer P_d fails, the transport of the message is interrupted because the outgoing edge no longer exists. The first message that can not be transmitted chooses their last node of the message path before entering the damaged peer P_d as restorer N_r. The restorer node organizes the reconstruction and signals, to all neighboring peers, to send undeliverable messages to it. All messages transferred to the restorer node have a marking whose edge (lost outgoing edge) of the message path is unavailable. So it can be said that the first node after the outgoing edge, i.e. after the transition, certainly belongs to the lost peer P_d. In the following, this node will be referred to as the transition node N_t and each diverted message has exactly one such transition node in its path $p_{N_0,N_x} = (N_0, \ldots, N_{t-1}, N_t, N_{t+1}, \ldots, N_x)$.

So the message path p_{N_0,N_x} can be split at N_t and all nodes of the partial path $p_{N_{t+1},N_x} = (N_{t+1}, \ldots, N_x)$ after N_t may or may not belong to the damaged peer P_d. The path could end on the damaged peer or the path could transition to another peer. A reconstruction of the nodes contained in the message path can lead to duplicates in the global subgraphs – a situation that should be avoided. All nodes of the partial path $p_{N_0,N_{t-1}} = (N_0, \ldots, N_{t-1})$ prior to N_t are certainly not on the lost peer P_d, otherwise the message would have been redirected earlier.

Considering the edges of a diverted message, the transition edge $E_t = (N_{t-1}, N_t)$ is an earlier outgoing edge and can safely be restored, because the message is undeliverable and was diverted to the restorer. Also the edge $E_{t+1} = (N_t, N_{t+1})$ belongs to the lost peer because it is associated with the transition node N_t. It can lead to a node within the peer or connect to another peer (outgoing edge).

In summary, a redirected message path is in a fixed order and can be split into partial paths as follows: The first $t-1$ nodes can definitely not be part of G_d (damaged peer P_d), followed by exactly one node N_t belonging to the graph G_d to be recovered. The assignment of the following nodes is uncertain (called "gray nodes").

3   Conceptional Approach

The most obvious way to cope with the failure of a single peer without data loss would be to store the data redundantly. This is how it is done in many areas of computer science. One of the best known is probably RAID-1 for the secure storage of data by storing the same data on several hard disks, as described in [3]. It would be easy to duplicate every global subgraph at least once on another peer. In the event of a failure, the duplicated peer simply takes over. But creating and keeping the copy consistent creates considerable effort, as shown in [6].

3.1   Experimental Setup

The idea is to use messages, which are created as a result of user queries containing valid path information, as a base for reconstruction. The queries follow on a power-law distribution, which means that some nodes are targeted often and others very rarely. This reflects the expected behavior of the message generation. Once created, the messages float through the remaining network as a normal payload of the P2P network and can be used without any additional (network) load or protocol. From the analysis of the containing path of these messages, which consists of concatenated nodes, it is possible to determine nodes that may or may not belong to the lost peer with certainty. Unfortunately, the messages itself do not provide enough information for most nodes to make this decision.

Is it possible to reconstruct the lost part of a distributed co-occurrence graph using only the knowledge gained from collecting and analyzing messages?

3.2 Basic Method

The algorithm used in this method is a slightly modified version of the algorithm presented in [6]. The authors are analyzing the diverted messages to find an outgoing edge in the message path *after* N_t. There are two cases to consider:

– If there is an outgoing edge, the message traverses the peer and the part between the node N_t to the outgoing edge is fully used for reconstruction. The following nodes are ignored because they are assumed to be associated with other peers.
– If no (known) outgoing edge is found, the assumption is made that the message ends on the lost peer P_d. The transition node and node N_{t+1} and its associated edges are restored. The remaining message path is ignored.

Table 1. Evaluation of message paths for 100k redirected messages

node of the remaining subpath after N_t	belongs to lost peer mean (std. deviation)
1. node (N_{t+1})	11.0243% (2.2957%)
2. node (N_{t+2})	1.1219% (0.7926%)
3. node (N_{t+3})	0.0927% (0.1234%)
4. node (N_{t+4})	0.0060% (0.0146%)
5. node (N_{t+5})	0.0004% (0.0031%)

Analysis of one thousand samples of 100k messages sent through a global graph (25 peers) with alternating lost peers

Especially at the beginning of the process, when only a few outgoing edges are known, the latter case is used often. Although only nodes N_t and N_{t+1} are added from the message path, after a few thousand messages analyzed, the resulting reconstruction contains an enormously large number of nodes that did not originally belong to the lost peer. An analysis of multiple samples of 100k messages sent through the global graph shows (Table 1) the reason for this. The messages extremely rarely end on the peer, but soon leave it over a (possibly still unknown) outgoing edge. Even if only the node N_{t+1} is used for reconstruction, in almost 90% (first row in Table 1) of the cases a wrong node is added in the process.

To mitigate the accumulation of false added nodes, the message part before the transition node N_t is used. These nodes cannot be part of the peer and are removed if they have already been allocated for rebuilding. When removing a false node, an outgoing edge to this node is added. This improves the processing of the following messages, since more outgoing edges are known, but removing superfluously created nodes with each message introduces another problem. At any time it can be determined that a node has not been part of the peer and therefore must be removed. However, the same node can be immediately added back again by the next analyzed message. Unlike the positively identified nodes, which are used to reconstruct the lost graph and are thus stored, the removal is not registered by any instance due to the stateless approach.

3.3   Cautious Approach

This simple and easy to implement method uses only assured knowledge. It uses the following elements of a message for reconstruction:

1. node N_t
2. the outgoing edge that leads to N_t
3. any edge E_i in the message path between two already reconstructed nodes is added

The algorithm prevents incorrect nodes or edges from being assigned for reconstruction at any time. The main disadvantage is that a significant proportion of nodes will never be restored. All nodes without outgoing edges, referred to below as core nodes (Fig. 2), never appear as N_t in the messages. In addition, the reconstruction is very slow because the nodes are not evenly distributed in the messages and some will hardly ever appear in a message. For these reasons, this method is not pursued further.

3.4   Exclusion List

The exclusion method attempts to solve the problem of the base method by providing a memory for nodes that could not have been part of the peer P_d. It takes the message path prior to the transition node N_t and stores its nodes in a data structure. All nodes after the node N_t are used for reconstruction if they are not in the exclusion list. If the exclusion list were perfectly filled, it would contain all nodes of all other peers except P_d. Even if this state is never reached, the size of the exclusion list tends to be large. The other, even worse, problem is that most nodes can not be identified as included or excluded at any time because they always appear after N_t in the message path.

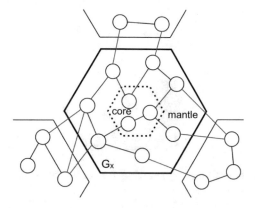

Fig. 2. Mantle nodes are all nodes with outgoing edges while core nodes just are connected to nodes within the peer

3.5   Shadow Effect

The promising simple approach of the exclusion list does not lead to success. The reason for this is that many nodes never appear before N_t or even as N_t. These nodes cannot be safely classified as included or excluded and prevent the exclusion list from growing further.

Some nodes always appear after N_t in the message path. A reason for this is that the nodes of the message path are ordered in a certain direction to flow through the global graph. The starting node N_0 has to be in the global subgraph G_i of the peer P_i ($N_0 \in L_i \cap G_i$). So all paths start in $L_i \cap G_i$ and additionally lie in a local graph L_i. If the selected local path happens to contain a mantle node of peer P_d, all nodes of the path following this node will never appear before the node N_t and thus cannot be classified. The situation is illustrated in Fig. 3. The hatched nodes N_1 - N_4 are also mantle nodes of P_d. All nodes that are in the message path behind this mantle node are in the shadow, because the mantle node will be the node N_t in the message analysis. If there is no other path leading to them, they remain hidden. The node N_6 is an exception. It is located behind a mantle node (N_4), but another message path through the node N_5 is possible to reach it and add the node N_6 to the exclusion list. Unfortunately, this is only useful in a few cases, since such messages are often not affected by the failure of P_d and are therefore not redirected to N_r, where they could be used for reconstruction.

In summary, the shadow effect leads to a considerable number of nodes that can never be categorized with certainty. All core nodes of P_d are particularly affected by this effect, as their edges are associated to mantle nodes or other core nodes of P_d. Therefore, all core nodes are always in the shadow of the mantle nodes.

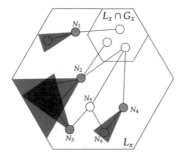

Fig. 3. The shadow effect when generating messages. The hatched nodes of L_i are mantle nodes of G_d

3.6   Improved Exclusion List

The improved exclusion list method tries a special approach to create a reliable allocation of the many nodes that cannot be categorized with certainty due to the shadow effect. In the following work, these non-assignable nodes (neither to the lost peer nor to another (still existing) peer) are referred to as gray nodes. The underlying idea is to use all the information currently collected (nodes and edges from the messages) to create a big graph and in order to sort the gray nodes. Figure 4 shows an example of what a graph created from all currently known nodes and edges might look like. The dashed line serves as a visual aid and categorizes the nodes. This help is of course not available during the reconstruction. The graph contains all known edges, many gray nodes and additionally nodes that already have been categorized.

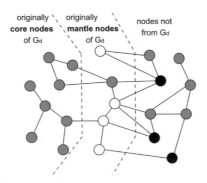

Fig. 4. Example of a graph created from all collected nodes and edges of all messages (gray: uncertain nodes; white: nodes certainly existed in the damaged peer; black: nodes which certainly are stored on another peer)

The next step uses the fact that the mantle nodes must have connections to both, core nodes and nodes on other peers. The nodes which are identified as stored on other peers have no direct connection to core nodes because core nodes are (by definition) just connected to nodes on same peer. The removal of all nodes reliably belonging to P_d is equivalent to the removal of all (known) mantle nodes of the lost peer. Since the mantle nodes form the boundary between P_d and the other functional peers, the remaining nodes are divided into different groups. Figure 5 illustrates the example graph after the removal of known mantle nodes and the resulting groups.

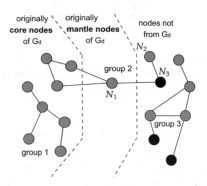

Fig. 5. Example of the graph from Fig. 4 reduced by the mantle nodes that have been identified with certainty

The resulting groups can now be distinguished by their proportion of (certainly) excluded nodes. Groups with a percentage below a threshold (e.g. 20%) can be assumed to contain nodes originally stored on P_d (e.g. 'group 1' shown in Fig. 5). These nodes will be restored. If the proportion is larger than the threshold (e.g. 'group 3' in Fig. 5), it is more likely that the previously unassignable nodes do not originate from P_d. These nodes are obviously ignored for reconstruction.

Afterwards, all nodes of the groups below the threshold and the known mantle nodes are restored. Figure 6 shows the result of the given example. The white hatched nodes were reconstructed, while the black hatched ones were discarded. Double lines show the outgoing edges of the reconstructed graph.

There remains a problem that still makes incorrect attribution possible. If the graph's mantle is not fully recovered, the removal of reliably contained nodes during the process is not complete. As a result, groups that should be separated remain together. To mitigate this problem, the threshold must be adjusted. When the threshold is set to zero, an excess node is rarely added because the group usually contains at least one excluded node, but it is still possible. This leads to a worse reconstruction. A higher threshold makes the method more stable against some non-recognized mantle nodes, thus increasing the reconstruction results at the cost of adding false nodes. Figure 5 shows this problem in group 2. The node N_1 was not recognized as a mantle node and was therefore not removed. Group 2 should actually be split, but is not. Since the threshold is not reached, all gray nodes of group 2 are restored. The erroneous addition of N_2 is

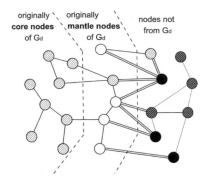

originally
core nodes
of G$_d$

originally
mantle nodes
of G$_d$

nodes not
from G$_d$

Fig. 6. The resulting example graph. The color of the shading of the nodes shows how they have been assigned: white for the nodes assigned by the algorithm to the lost peer, black for those assumed to exist on other peers.

accepted, otherwise the whole group would have to be discarded. If the mantle nodes are fully identified, this problem no longer occurs.

For the duration of the reconstruction, the presented method must cache all potentially relevant data, temporarily reconstruct the entire global graph, and perform computations on it. Both storage and computational overhead can therefore be significant for this method. A possible implementation of the improved exclusion list method is shown in Algorithm 1 and 2. Here, a "slightly" more efficient approach is shown, avoiding forming and computing on subgraphs. Nodes are sorted into sets (called buckets) based on their associated edges. Nevertheless, before reconstruction, all information given by messages must be collected and stored.

The algorithm therefore uses two data structures: one for storing the node data ("node_list", see Table 2) and another for the edges ("edge_list"). The latter consists of a list of two node IDs connected by a respective edge. For the Algorithm 1, it is assumed that no duplicates are created in the data structure, even with multiple insertions.

Table 2. Description of an element of the node list

name	type	description	size
id	integer	shorter than 'content'	depends on list size
content	string	word in co-occurrence graph	various
state	integer	'inside', 'outside', 'unknown'	2 Bit
bucket	integer	id of the bucket	like id

After the mapping process (Algorithm 2), all nodes of a group are in the same bucket. If the percentage of foreign nodes (located on other intact peers) is

---

Algorithm 1 process message

Require: *node_list, edge_list*

1: **function** PROCESSMESSAGE(*m*)
2:    **for** *node* in *m.path* **do**
3:       **if** *node* in *node_list* **then**   ▷ node already exists
4:          *stored_node* ← *node_list.get*(*node*)
5:          **if** *node.state* ≠ 'unknown' **then**
6:             **if** *node.state* ≠ *stored_node.state* **then**   ▷ update state
7:                *node_list*[*stored_node*].*state* ← *node.state*
8:             **end if**
9:          **end if**
10:       **else**   ▷ new node
11:          node_id ← node_list.add(node)
12:          **if** node.state ≠ 'mantle' **then**
13:             node_list[node].bucket ← node_id   ▷ initialize bucket
14:          **end if**
15:       **end if**
16:    **end for**
17:    **for** edge in m.path **do**   ▷ process edges in message path
18:       s_id ← node_list.get_id(edge.s)
19:       t_id ← node_list.get_id(edge.t)
20:       edge_list.add(s_id, t_id)
21:    **end for**
22: **end function**

Algorithm 2 group mapping

Require: *node_list, edge_list*

1: **for** *edge* in *edge_list* **do**
2:    ▷ use edge if no mantle node is contained
3:    state_s ← node_list[edge.s].state
4:    state_t ← node_list[edge.t].state
5:    **if** state_s ≠ mantle & state_t ≠ mantle **then**
6:       ▷ get current buckets from node list
7:       bucket_s ← node_list[edge.s].bucket
8:       bucket_t ← node_list[edge.t].bucket
9:       **if** bucket_s ≠ bucket_t **then**   ▷ if buckets differ
10:          **for** node in node_list **do**   ▷ merge buckets
11:             **if** node.bucket = bucket_t **then**
12:                node_list[node].bucket ← bucket_s
13:             **end if**
14:          **end for**
15:       **end if**
16:    **end if**
17: **end for**

---

below a threshold, all nodes of this bucket can be restored along with the found mantle nodes.

Once all relevant information is found, storage consumption (*memory*) can be estimated as follows:

$$\left(1 - \frac{1}{|P|}\right) \cdot \kappa \leq memory \leq |P| \cdot \kappa$$

with $|P|$ as the number of participating peers and $\kappa = s_n \cdot n_{max} + s_e \cdot e_{max}$ where s_n (s_e) is the (average) storage size in bytes of an element.

The maximum number of nodes $n_{max} = \lceil |n_k| \rceil, k \in 0, 1, .., |P| - 1$ (edges $e_{max} = \lceil |e_i| \rceil, i \in 0, 1, .., |P| - 1$) found on any peer of the network in the intact global graph. The storage requirements of the exclusion list algorithm fall between the upper bound representing storing the entire global graph and the lower bound in case the peer that hosted the most nodes and edges fails and no redundant data is found (in this case, no reconstruction is possible anyway).

3.7   Mantle-Before-Core

To preserve the advantages of improved exclusion list method and reducing the offside of computational and memory consumption, an alternative approach based on two phases was developed.

1. **Phase 1:** Collect only nodes which were safely located on the damaged peer (all nodes N_t).
2. **Phase 2:** In addition, all nodes in the path of a message that lie between two known mantle nodes are added.

The main goal in phase 1 is to restore as many mantle nodes of the damaged peer as possible. The reconstruction is performed in this phase without incorrectly added nodes, but also without restoring a single core node. Assuming that the mantle is fully recovered, each message path traversing the core can now be identified. Such a message path must contain another known mantle node after the initial passage of N_t. So it can be taken for granted that all nodes between these two mantle nodes are part of the core. Phase 2 is about reconstructing the core, which was completely disregarded in phase 1. Two critical points remain:

1. If the mantle is not fully recovered - which will almost always be the case due to lack of redundancy - a path can always break out "undetected", resulting in too many nodes being restored.
2. The point in time at which to switch over has to be chosen wisely, although there is insufficient local information available for this decision.

The second point influences the first: if the switch is made too early, too much of the cladding is not restored, which will inevitably lead to a large proportion of the nodes being over-added. Switching too late delays reconstruction and prevents early use of all information given by the received messages. This problem is exacerbated by the fact that the destination nodes of the messages are not evenly distributed, but follow a power law distribution. Thus, some of the nodes appear disproportionately often, while others are almost never part of a message. In return, this method does not require the incoming messages to be stored for later processing. The profitable linking of the message contents takes place exclusively in the reconstructed graph. This significantly simplifies the necessary calculations compared to the improved exclusion list.

4   Experiments and Results

The experiments conducted did not simulate a real P2P network, but only the generation and processing of messages.

25 local graphs each with about 1900 nodes and 6000 edges in average were used. Each local graph consists exactly of one fully connected component. All local and global graphs satisfy the small world properties [7]. The simulation consisted of 100 runs, with 15 of the 25 local graphs randomly selected for each run. In the next step, a global graph was constructed from the local graphs, which in turn was distributed to 15 peers and one random peer was removed again. Thus finally the P2P network is missing a local graph and a part of the global graph (which is to be reconstructed). A peculiarity of our underlying data (local graphs) is that about 60% of the nodes are stored in only one local graph. Therefore, it is possible that these nodes cannot be reconstructed if they also happen to be stored in the global graph of the failed peer. Some nodes are therefore irretrievably lost. For this reason, the following diagrams show the reconstruction rate related to the number of recoverable nodes of G_d and not to all nodes that were originally in G_d. This type of counting makes the actual capability of the algorithm in question clearer, since a greater or lesser degree of reconstructability by chance does not matter. Only the messages which are used for reconstruction are shown in the diagrams. Since most messages traverse the global graph without noticing an error, this is a subset of all generated messages. In our experiments with 15 peers, about every fifth message was involved in the reconstruction. The main challenge is not to restore a large part of the lost nodes, but to prevent large numbers of surplus nodes from being created.

4.1   Basic Method

The performance of the basic method in terms of the percentage of nodes recovered is shown in Fig. 7. After 100,000 messages, the reconstruction of correctly added nodes is mostly complete and there are no further significant improvements. On the other hand, the surplus of added nodes increases steadily over the entire time.

The naive approach of simply stopping the procedure at about two hundred thousand messages, thus ending the reconstruction with a perhaps tolerable level of surplus added nodes, assumes that this limit (here: 200,000 messages) can be determined with sufficient precision. But this message count is only valid in our experimental setup and might change with the number of peers, the sizes of the local graphs, the redundancies present, and so on. Using a static value, one would then risk either adding too many surplus nodes or stopping too early and thus achieving an even lower reconstruction rate than is even possible with this method.

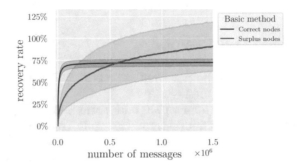

Fig. 7. Simulation results of the basic method

On average, the recovery rate only reaches about 70% of the possible reconstructable nodes, which is not particularly good compared to the other methods.

4.2   Exclusion List

The exclusion list method (Fig. 8) achieves a much higher recovery rate (over 90%) and reaches a point with no further improvements at around 100,000 processed messages. This reconstruction, which is significantly better than the basic method, comes at a high price: The surplus added nodes reaches values above 800% very early on. A graph reconstructed in this way contains de facto all recoverable nodes, but also an approximately tenfold higher number of nodes that already exist on other peers. This happens because this method adds all nodes whose assignment cannot be excluded with certainty.

4.3   Improved Exclusion List

The improved exclusion list performs, in terms of correctly reconstructed nodes, practically as well as the normal exclusion list. The biggest difference are the surplus nodes. After a short increase within the first thousand messages, the number of incorrectly added nodes decreases and stabilizes below 10%. This effect exists because as more mantle nodes are found, the division into groups becomes more error-free (Fig. 9). Our worries about the storage requirements were confirmed in the simulation. Figure 10 shows that very early on, more than 80% of the nodes originally present in the global graph must be stored before reconstruction. After about one million messages, more than 90% of the nodes

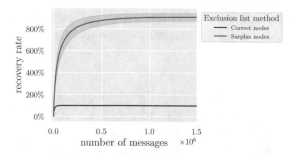

Fig. 8. Simulation results of the exclusion list method

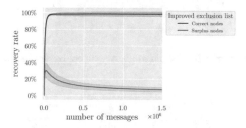

Fig. 9. Simulation results of the improved exclusion list method

from G have already been cached for subsequent reconstruction. Since this storage is done on a single peer, it is a major challenge. The effort for the necessary calculations on the large amount of data is then also correspondingly large.

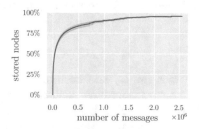

Fig. 10. Simulation results of memory usage of the improved exclusion list method; number of stored nodes in relation to number of all nodes in the global graph G

4.4 Mantle-Before-Core

As expected, this method achieves minimal restoration in phase 1. The detailed image in Fig. 12 shows this phase, in which only the nodes that are certain to come from the damaged peer P_d are added.

After switching to phase 2, the correctly added nodes increase rapidly, up to a nearly complete restoration of all recoverable nodes. At the same time, the number of false nodes increases rapidly. Unfortunately, these nodes do not disappear over time, but find a balance between adding and removal, as shown in Fig. 11.

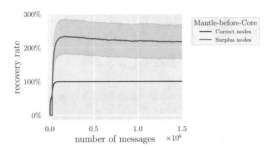

Fig. 11. Overview: Nodes correctly restored by Mantle-before-Core

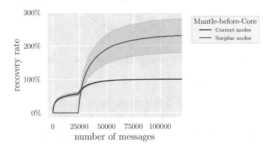

Fig. 12. Detail: Nodes correctly restored by Mantle-before-Core

4.5 Comparison

In terms of restoring correct nodes, the stateful approaches (both exclusion lists) and the mantle-before-core approach achieve almost complete restoration of the global subgraph (Fig. 13). The basic method finds significantly fewer correct nodes. Looking at the surplus nodes added, a clear ranking can be found (Fig. 14). The exclusion list method performs disastrously, followed by the Mantle-before-Core method, which only adds about every second node too much and ends in an almost constant progression. The basic method starts at an acceptably low level, but keeps adding surplus nodes over time. The improved exclusion list performs best in terms of surplus nodes added.

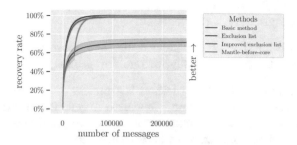

Fig. 13. Comparison: Correct restored nodes

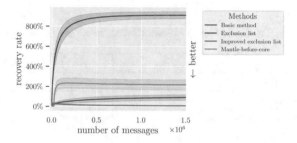

Fig. 14. Comparison: Surplus restored nodes

5   Discussion

If each message is considered individually, a significant portion of the information required for reconstruction is ignored because the nodes contained in the message path cannot be classified. Splitting the message path at the transition node N_t and assigning many subsequent nodes after N_t for reconstruction results in a very high number of incorrectly assigned nodes. Their growth is so high that the correction by the nodes located in the message path before the transition node is too weak to dampen this growth.

For this reason, a lean, stateless approach, where each message is analyzed individually, leads to a less good result. This is shown by the results of the experiments for the base method and Mantle-before-Core method (especially after the switch). Accumulating information from multiple messages and thus using a stateful approach is advantageous because more unclassifiable nodes (gray nodes) can be categorized. This is especially true when using the improved exclusion list method, where groups of nodes are formed in order to determine their association. After all, the improved exclusion list method has some clear advantages:

– a little faster reconstruction
– (much) less surplus nodes

But there is a price for these benefits:

- storage needed for storing all collected message information
- expensive calculations (especially for building groups)

Since the Mantle-before-Core method is stateless, it is much more resource efficient, but its reconstruction result is worse. A generally valid answer to the question of the switchover time remains open, even if good values can be determined for certain situations. However, the results are so good that the use of the more complex, improved exclusion list procedure must be weighed up depending on the general conditions (compare Table 3).

Table 3. Comparison of introduced methods

	Correct Nodes	Surplus Nodes	Remark
Basic method	- -	−	
Exclusion list	++	- -	expensive
Improved exclusion list	++	++	expensive
Mantle-before-Core	++	−	

If the cost of removing the (sometimes quite large number of) superfluous nodes after completion of the reconstruction is small, then the expense of a stateful procedure is probably not worthwhile. If, on the contrary, the goal is to achieve an error-free (redundancy-free) reconstruction, the improved exclusion list is undoubtedly the better choice. At this point, protocol extensions that are conceivable at various points were completely disregarded, even though they can achieve a significant improvement in the reconstruction with little additional effort, depending on the case. Of course, all such approaches contradict the paradigm of cost-neutral reconstruction with regard to network utilization.

6   Future Work

Various aspects can still be examined in the future. First, the switchover point of the mantle-before-core is a key success factor and in the simulations carried out so far has only been determined specifically for the simulated network. A more general algorithm suitable for different networks would be indispensable for practical use.

Beyond that, however, there are at least two other very interesting areas. On the one hand, the algorithms found should be tested for their suitability in more complex simulations. This includes not least the behavior in case of failure of several peers, failures during the reconstruction process and a variation of the redundancy in the local graphs. On the other hand, extensions of the protocol are conceivable, which could have far-reaching effects. For example, the remaining peers can not only continue to send messages, but also selectively participate in the reconstruction by intentionally sending specific information necessary for the reconstruction. Of course, this is always at the expense of the load in the network, but a cost-benefit analysis would be necessary here if the reconstruction could thus be significantly accelerated or run more error-free.

7 Conclusion

In the situation without an enforced redundancy of a peer the reconstruction will nearly never be perfect. Due to the loss of a local graph and part of the global graph, nodes and edges that were stored *only* on the lost peer cannot be recovered, regardless of the algorithm used. Nevertheless, it has been shown that there are at least two practicable ways to introduce fault tolerance into this particular peer-to-peer network, using solely the messages that flow through the network by default. Each variant leads to very good reconstruction results, but both approaches also have disadvantages. Either one decides for many superfluous nodes (which have to be removed again) or for a higher consumption of memory and computing resources during the reconstruction time.

In conclusion, it is possible in this particular network to introduce and establish procedures that can start immediate regeneration and compensation in case of a partial failure, using only the remaining information of the network. In this way, global knowledge is regained that at first glance seems lost. However, full recovery of a peer is not possible due to the incomplete redundancy of the network (e.g. the peer's local graph remains lost). At least the global graph and thus the global knowledge gets a "free" recovery option.

References

1. Almeida, S., Vicentini, J., Bonilha, L., De Campos, B., Casseb, R., Min, L.: Brain connectivity and functional recovery in patients with ischemic stroke. J. Neuroimaging **27**, 65–70 (2017). https://onlinelibrary.wiley.com/doi/abs/10.1111/jon.12362
2. Kubek, M., Unger, H.: The WebEngine: a fully integrated, decentralised web search engine. In: Proceedings of the 2nd International Conference on Natural Language Processing and Information Retrieval, pp. 26–31 (2018). https://doi.org/10.1145/3278293.3278294/
3. Patterson, D., Gibson, G., Katz, R.: A case for redundant arrays of inexpensive disks (RAID). In: Proceedings of the 1988 ACM SIGMOD International Conference on Management of Data, pp. 109–116 (1988). https://doi.org/10.1145/50202.50214
4. Simcharoen, S., Unger, H.: The brain: WebEngine version 2.0. In: Unger, H., Kubek, M. (eds.) The Autonomous Web, pp. 51–68. Springer, Cham (2022). https://doi.org/10.1007/978-3-030-90936-9_4
5. Simcharoen, S., Nagy, G., Unger, H.: Decentralised routing in P2P-systems with incomplete knowledge. In: Unger, H., Kubek, M. (eds.) The Autonomous Web. Studies in Big Data, vol. 101, pp. 123–135. Springer, Cham (2022). https://doi.org/10.1007/978-3-030-90936-9_9

6. Unger, H., Kubek M., Sukjit, P.: Netzbasierte Ansätze zur natürlichsprachlichen Informationsverarbeitung. Springer Fachmedien Wiesbaden (2022). https://doi.org/10.1007/978-3-658-37284-2
7. Watts, D., Strogatz, S.: Collective dynamics of 'small-world' networks. Nature **393**, 440–442 (1998). https://www.nature.com/articles/30918

Implementing the WebMap: An Extension to the Web

Georg Philipp Roßrucker[✉]

Communication Networks, FernUniversiät in Hagen, Unistr. 1, 58084 Hagen,
Germany
g.rossrucker@gmail.com

Abstract. The WebMap is an advancement of the previous *WebEngines 1* and *2*. It facilitates a distributed extended linking structure for any hyperlinked network of documents like the Web. It is designed as a modular system in which each participant can assume different roles and functions. The current progress of the implementation as well as the results of a first experiment in a single host setup are presented. They indicate the feasibility of the concept and suggest further tests in multi-host environments and larger hypertext-based networks. Furthermore, third parties can utilize the WebMap and develop extended services based on it. In here, the example of a web search application is showcased.

1 Introduction

Building a co-occurrence graph based on words that often appear together in a given text coprus allows to build a network of words that assumably are contextually close to eachother. [1] showed that the information stored in such a graph can be utilized to derive *text representing centroids* (TRC) for any kind of input text. This representing term can be interpreted as an contextual center of gravity for the given input.

[2] showed how to construct a decentralized web search engine utilizing co-occurrence graphs and TRCs: They created a global co-occurrence graph by joining the local graphs of multiple hosts into one global graph and distributed it among all participants. Documents were then assigned to the global graph's nodes according to their TRC. Searching for documents was then achived by deriving a TRC for a search query and determining a route towards the TRC's node representation on the global graph. Finally, all documents assigned to that node were returned as a search result to the user.

Based on this idea, a concept was introduced in [3], which intends to improve the usability and accessibility of the distributed web engine by introducing static cluster files. Cluster files represent the former global graph nodes and provide the associated information in commonly accessible formats and protocols. Hyperlinks form the edges of this graph which can be placed on the web as an extension to the existing web graph, which is referred to as "WebMap".

In this report the current progress in developing this decentralized static linking structure is described. Also an initial experiment to demonstrate the

H. Unger and M. Schaible (Eds.): Real-Time 2022, LNNS 674, pp. 171–184, 2023.
https://doi.org/10.1007/978-3-031-32700-1_17

feasibility was conducted and the results are presented here. Furthermore, a DNS-like lookup system for clusters and cluster hosts was implemented and utilized in the setup as an alternative approach to crawling. The details of the implementation are presented in Sect. 3, the experimental setup in Sect. 4, and the results and discussion in Sect. 5.

2   Idea

The basic working principle of the WebMap is to cluster documents by a common contextual term as drawn from [2]. In contrast, these clusters are represented by text documents which are directly placed on the web instead of being kept in local graph databases, which are only accessible to the application server and connected nodes of the underlying P2P network. Clusters that have a certain contextual proximity are linked to each other, so that in steady-state a linked network of clusters emerges. This network can then be utilized by any consumer on the web to provide extended web services such as web search or others.

The goal of the current prototype implementation is to provide a modular system in which the participants can assume any of several roles required to create a global network of linked clusters. Further, a cluster look up system (CLS) is introduced which helps to identify and locate other clusters on the web.

Cluster Network. In contrast to the global graph in the former approach the cluster network represents a *real* extension to the web graph. Instead of being only accessible to participating peers, clusters are text files stored in a standardized location on cluster-servers and are accessible to anyone on the web. They contain hyperlinks to related documents and other related clusters. The contextually linked cluster files form a distributed graph network. In analogy to site maps for single websites this cluster network is referred to as the "WebMap".

Cluster Lookup System. The cluster look up system is introduced as a distributed lookup system resolving cluster names to their hosts' names. This helps to derive the host of a given cluster in a more efficient way than blindly crawling the web for cluster hosts. However, it requires that participating cluster hosts subscribe to a CLS server.

Roles and Participants

- **Web servers** provide documents and assign them to clusters utilizing a local co-occurrence graph. Furthermore, they provide a common API to derive information from their local graphs, such as TRCs or shortest pathes between nodes.
- **Cluster hosts** provide cluster files in a standardized location to the web. They provide a common API to web servers through which related documents or clusters can be assigned to or revoked from a cluster.

- **CLS servers** provide an API to request the hostname of a given cluster, or the clusters of a given hostname. They also provide an interface to subscribe and unsubscribe cluster hosts, and retrieve their clusters regularly to update a local cluster-host index.
- **Consumers** access the information provided by the other roles through their APIs and can build additional services such as search engines, recommender systems, or others.

Any participating host of the WebMap can assume any of the roles at the same time. The WebMap functions as those roles interact and provide functionality to each other. Table 1 presents an overview of possible interactions.

Table 1. Roles and relevant functionality provided to eachother. 1st column: providing role, 1st row: receiving role

	Web server	Cluster Server	CLS Server	Consumers
CLS Server	Interface to derive a cluster host		Interface to forming a P2P-network	Interface to derive a cluster's host, or all clusters per host
Cluster Server	Interface to add or remove documents and pathes of related clusters	Interface to negotiate the cluster authority	Interface to derive a list of local clusters	Interface to derive a cluster
Web server	Documents and interface to access the local graph			
Consumer	Additional functionality/services			

3 Implementation

In accordance with the allocation of roles and functions, the prototype implementation follows a modular approach, in which each of the roles presented above can be provided by any participant.

3.1 Cluster Lookup Server

The hosts acting as Cluster Lookup Servers are part of the Cluster Lookup System (CLS Server). Each CLS Server maintains a list of cluster hosts subscribed to it. Periodically, it queries the subscribed cluster hosts for their clusters in order to maintain a local directory that maps all clusters to their cluster servers.

The CLS server provides an API to allow external inquirers to:

- Request the host of a given cluster.
- Request all clusters of a given host.

In this prototype implementation only a single CLS Server is provided. It is introduced to all participating cluster servers and web servers. In a productive implementation the Cluster Lookup System should be set up in a distributed and self-organized manner. For this purpose a P2P-network, e.g., a Chord-like [4] structure could be implemented. Here the CLS information would be distributed across the participants and could be looked up by all internal and external inquirers.

3.2   Cluster Server

A participating host who acts as cluster server maintains and provides cluster files to the web. The following functions are provided by the cluster server:

- A list of the clusters managed by the host.
- The content of any of the managed clusters.
- The addition and removal of related documents to/from a cluster.
- The creation of a hyperlinked path of related clusters from a starting towards a target cluster (and the creation of missing clusters on this path).
- The addition and removal of related clusters to/from a cluster.
- The negotiation of a cluster authority with another cluster server.

These functions can be requested via an HTTP API. The hosted clusters are also accessible via an standard HTTP web server. To register the cluster server to the cluster lookup system it requires an CLS server as an input.

3.3   Web Server

A participant assuming the role of a web server provides a standard HTTP web server through which it hosts documents to the connected network, e.g. the web. In addition, it maintains a local co-occurrence graph, which allows it to determine representing terms for its local documents, and later on enables the creation of pathes and routing across the cluster network. The procedured regarding the construction and preparation of co-occurence graph implemented and described below follow the approaches presented in [1]. The web server requires the following inputs:

- The host serving as default cluster server to the web server.
- The minimum number of documents required to provide a mature co-occurrence graph and to allow computation of TRCs and routes.
- The minimum distance for edges to remain on the co-occurrence graph.

With these inputs the web server can now perform the following tasks:

Building the Local Graph: While the number of local documents is smaller than the minimum number required for a mature graph, the following is repeated for each new local document:

1. Pre-process the document to derive a set of normalized words, independent of grammar and language-specific stopwords [5,6]: Split text by sentences, remove numbers, remove punctuation, remove stopwords, lemmatize words.
2. Add the remaining, normalized words as nodes to the local co-occurrence graph and link all words that appear in the same sentence. Add the numbers of occurrence to the nodes and co-occurrences to the edges.

Preparing the Local Graph: Once the co-occurrence graph has achieved a mature state, i.e., the minimum number of documents has been reached, it needs to be optimized in order to efficiently derive TRCs and routes from it. Optimization means removing weak links, which is achieved by the following steps:

1. For each edge, compute the dice-coefficient [7] based on occurrences and co-occurrences of the adjacent nodes. It can be interpreted as a measure of proximity. Its inverse represents the distance of the nodes.
2. Remove all edges who's distance drop bellow the minimum distance from the co-occurrence graph.
3. Remove all isolated nodes from the graph.

Deriving a TRC per Document: For the subsequent assignment to a cluster on the WebMap, each document needs to be represented by a single term. For this, the TRC of each document is determined, utilizing the local co-occurrence graph. TRCs are derived by implementing the spreading activation technique introduced in [8]. To optimize TRC computation per document it is executed per sentence and in batches as follows:

1. Compute the TRC for each sentence of a document. Only words that are connected on the graph are considered.
2. Iteratively derive TRCs for the sentence-TRCs in batches of 10 until one last TRC is left.
3. The last remaining TRC serves as the designated document TRC.

Adding Documents to Clusters and Creating the Cluster Network: For each document for which a representing term was derived, the following steps are conducted in order to add the document to the global cluster network:

Verify, if the default cluster server maintains the respective cluster.
If yes: Request the cluster server to add the document to the cluster.
If not:
(a) Derive the shortest path from an existing cluster towards the target cluster from the local co-occurrence graph.
(b) Request the cluster server to derive and if necessary create a hyperlinked path towards the target cluster and to add the document to this cluster.

The cluster server acts as requested and either adds the document to the respective local cluster or creates a hyperlinkned path from a local cluster to the target cluster by executing the following steps for all clusters on the path:

1. Request the host of the *next cluster* from the CLS Server.
2. If the *next cluster* does not exist: Create it locally.
3. Request the next cluster's host to add the *current cluster* as a related cluster to the *next cluster*.
4. Request the current cluster's host to add the *next cluster* as a related cluster to the *current cluster*.

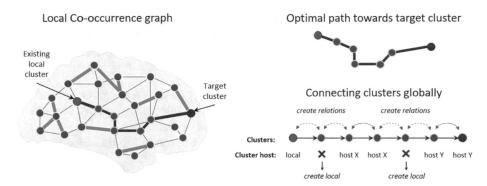

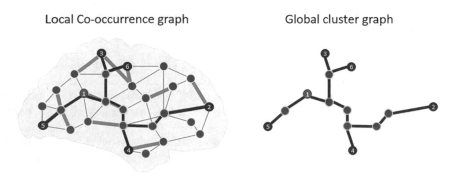

Fig. 1. Connecting clusters globally by deriving a path from the local co-occurrence graph.

5. If the *next cluster* is the *target cluster*: Request the cluster's host to add the document as a related document to the *next (target) cluster*.

Figure 1 illustrates the process of deriving an optimal path from a known local cluster towards the desired target cluster from the local co-occurrence graph and creating a path of mutually linked cluster files on the web. Missing clusters on the path are created by the default cluster server. Repeating this for all documents results in a global graph of connected clusters which form a subgraph of the local co-occurrence graph. Figure 2 depicts this for the case of a single co-occurrence graph. When this is jointly performed by multiple web servers, the resulting global cluster graph is a merger of all sub graphs of all local co-occurrence graphs, representing the optimal pathes of all local graphs.

Fig. 2. The global cluster graph represents an optimal sub graph of the local graph.

3.4   Search Service

The initial idea was to adapt the search approach presented in [2]: Based on a local co-occurrence graph the search service derives a path from any cluster whos location is known towards the target cluster. It is based on the assumption that all participating peers' local co-occurrence graphs are projected into the global graph and *every path* derived from *any local graph* must therefore be traceable on the global graph, too. As described earlier, in the proposed implementation, it is not necessarily the case that all nodes and edges of all local graphs are reflected by the global graph. Even though in steady-state it is expected that all nodes have a cluster representation, it cannot be assured that all edges of all local graphs exist globally.

One way to address this would be to implement an algorithm that allows to approximate towards the targeted TRC cluster or any closest alternative on the global graph. Depending on the size of the graph, scalability needs to be addressed by such an algorithm since it may require extensive computing power and time to do so. Instead, it is proposed that a search index of the existing clusters is maintained which would be accessible to the search service and returns the desired output. The following two options are suitable:

- The search service provider periodically crawls the network of connected cluster files and maintains its own index of clusters and hosts.
- The search service provider utilizes the cluster lookup system as an index to retrieve clusters and hosts.

From this index the search service provider retrieves the host of the intended cluster and can directly access it to derive related documents. In the prototype implementation the existing CLS is utilized for this purpose. Besides the CLS server the search service takes the web server, whose local co-occurrence graph shall be utilized, and a user generated search query as an input for a search request. Then the following steps are conducted to retrieve search results:

1. Normalize the query, i.e., remove numbers, remove punctuation, remove stopwords, lemmatize words.
2. Derive a TRC for the normalized query from the co-occurrence graph of the associated web server.
3. From the search index (CLS Server) check if a cluster for the given term exists and derive its host.
4. Access the cluster and derive the related documents from it.
5. Return the search results to the user.

In addition the search results can be extended by applying any of the following methods:

- Compute alternative TRCs, which means second and lower ranking TRC candidates, and derive the related documents from their associated global clusters.
- From the local co-occurence graph get the neighbors of the TRC node and derive related documents from their associated global clusters.
- From the TRC's (global) cluster derive related clusters and their related documents.

All of these methods are based on the assumption that associated clusters also contain contextually related documents and may serve as a suitable extension to the search results.

4   Experimental Setup

4.1   Setup and Data

This initial experiment serves as a proof of concept and shall demonstrate functionality and feasibility of the WebMap and a web search application. This means that it is intended to demonstrate the construction of the network of connected cluster files, to show that any document can be added to this network and is trackable when an associated search query is entered into the search service operating on it.

The dataset used for the experimental setup is the dataset 1.2 drawn from [9]. It consists of 200 plain-text documents of German news articles. 50 documents for each of the topics "finance", "sports", "cars", and "politics".

To keep the setup as simple as possible, the prototype implementation is deployed on a single host only, which combines all roles and functions described above. This is an important constraint, especially for the cluster server and the CLS server which are both not distributed across multiple hosts and therefore, several aspects necessary to be addressed in distributed systems like load balancing, precautions for possible outages of remote hosts, and others are not considered and implemented here.

4.2   Creating the Local Co-occurrence Graph and Deriving Document TRCs

Within the web server component all 200 documents were pre-processed, resulting in a total of 5,959 sentences and 13,576 unique words. These words were then added to the local co-occurrence graph, and connected, according to their joint appearance within sentences resulting in 189,744 edges. To improve the performance of the subsequent computation of TRCs per documents weak edges were removed. The strength of edges was determined by deriving the distances of the adjacent nodes according to Sect. 3.3.

In the first step the graph was adjusted for edges who's distance was greater or equal the median distance resulting in a remainder of 93,557 edges. As a result also 73 isolated nodes were removed from the graph.

To further emphasize the heavy weighted edges, the same procedure was applied again, so that another set of edges whose distances were greater or equal to the median distance of the remaining set of edges were removed. Doing so resulted in 45,770 edges which is approximately 25% of edges which were initially present. 833 isolated nodes were lost. Applying the same again resulted in even larger shares of isolated and removed nodes. Table 2 presents the details for 5 iterations after which the graph was entirely cleared.

Besides isolating single nodes, reducing weak edges can also lead to larger disjoint components of the graph. Therefore, a trade-off between performance and a well connected graph needs to be found. Testing the graph after each iterations showed that already after the second iteration larger disjoint components appeared. Since the performance after one iteration was already acceptable, further experiments were conducted based on the graph that resulted from the first round of adjusting for weak edges. Other methods to evenly remove weak edges while keeping the graph well-connected should be evaluated and tested subsequently.

Table 2. Graph statistics after initial creation and five iterations of adjusting for weak edges and isolated nodes.

Iteration	Nodes	Edges	Median Dist.	Mean Dist.	Max Dist.
0	13,576	189,744	11.0	20.07	$< \infty$
1	13,503	93,557	4.5	4.84	< 11.0
2	12,743	45,770	2.25	2.33	< 4.5
3	10,720	22,850	1.5	1.45	< 2.25
4	6,819	9,135	1.0	1.01	< 1.5
5	0	0	-	-	-

The next step in setting up the Web Map was to utilize the co-occurence graph and compute a TRC for each document. With the prepared graph and the procedure described in Sect. 3.3 this resulted in 168 different TRCs for the 200 mixed-topic document.

4.3   Creating a Global Network of Clusters

Next, the procedure of deriving shortest pathes from the co-occurrence graph for all the TRCs' associated clusters and existing clusters was conducted iteratively. In each iteration this route was derived and converted into a hyperlinked path of cluster files and the reference (URL) of the respective document was attached to the target cluster.

The final network of linked cluster files represents a sub-graph and a spanning tree of the local co-occurrence graph. It may not be a minimum spanning tree, as edges are created depending on the order in which nodes, i.e., documents, were added and not with the premise of forming a *minimum* spanning tree. The network consists of a total of 258 clusters with 257 hyperlinked edges among them. 168 of these clusters have documents assigned to them. With this setup the first prototype of a single-host, non-distributed WebMap was created. Figure 3 illustrates this network.

4.4   Searching for Documents

Since a query TRC is computed locally, for any given input query the query TRC must be an element of the set of nodes of the local co-occurrence graph. In this one-host setup the search for related documents can therefore be simulated by determining the yield of search results for any node of this graph. From the construction of the graph it follows that adjacent nodes have a contextual similarity. Therefore, adjacent nodes can be considered to extend the search scope. To test this, the following was conducted:

- Determine the number and share of nodes for which no results, i.e., no documents, are found.
- Derive the numbers of related documents for each node from their associated clusters.
- Derive the same numbers for each node, taking adjacent nodes into account.

Global cluster graph

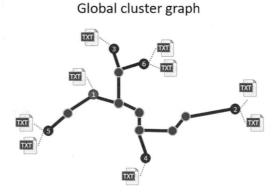

Fig. 3. The global cluster graph with documents attached to the nodes.

In this setup the results will be the same regardless of choosing neighbors from the local graph or from the global cluster graph. In a distributed WebMap both approaches may result in different search results. The latter may enrich search results by taking clusters into account that are not nodes of the searching host's co-occurrence graph.

The simulation was performed in three series. The first did not take neighboring nodes into account while the second took the 1st degree neighbors and the third also the 2nd degree neighbors of each graph node into account. Table 3 depicts the results of these test series.

Table 3. Results of the three runs of the experiment. Showing the share of nodes for which documents were found and the number of documents found per node of the local co-occurrence graph.

Number of nodes on the local co-occurrence graph: 13,503

	Documents found (no neighbors taken into account)	Documents found (1st degree neighbors taken into account)	Documents found (2nd degree neighbors taken into account)
Number of nodes for which no clusters were found	13,245	10,122 *(including 1st degree neighbors)*	1,283 *(including 2nd degree neighbors)*
Number of nodes for which a cluster was found but no related documents / *Number of nodes for which clusters were found but no related documents*	90	448	279
% of Nodes for which documents were found	1.24 %	21.72 %	88.43 %

Number of related documents identified	Nodes of the local graph (no neighbors)	Nodes of the local graph (1st degree)	Nodes of the local graph (2nd degree)
1	147	1,751	1,350
2	17	613	1,281
3	3	208	1,068
4	0	139	928
5	0	98	807
6	0	46	772
7	0	27	600
8	0	8	592
9	0	2	426
10	1	1	409
11-20		11	2,100
21-30		28	913
31-40		1	446
41-50			179
51-60			55
61-70			13
71-80			2

Average number of search results per node			
only nodes with results	1.19 (168)	2.08 (2,933)	9.82 (11,941)
all nodes	0.015	0.32	8.68

5   Results and Discussion

From the experimental setup described above some basic observations can be drawn. The 200 documents were assigned to 168 different clusters according to ther document TRCs. A total number of 258 clusters were created in this single-host setup, which in this special case form a spanning tree without any loops. This represents only a minor share of the possible clusters according to the local co-occurrence graph, which comprises of 13,503 nodes after one iteration of removing weak edges and isolated nodes. How the global graph evolves with a rising number of participants and documents needs to be tested.

With regard to the search simulation, the yield of the search, as number of results per possible query TRC was examined. As a conclusion it can be stated that taking neighboring nodes into account makes a remarkable difference in the number of documents found per node and the share of nodes for which documents were derived at all. In the case that no neighbors were taken into account, only 1.24% of nodes (168) lead to existing clusters, with documents attached to them. Of the remaining 13,335 nodes, the majority of 99.3% did not even have an existing cluster representation.

Taking the 1st degree of adjacent nodes into account increased the number of nodes, which produced any search result to 21.72%. When taking the 2nd degree neighbors into account this number even rose to 88.43% of nodes, meaning that for the majority of possible query TRCs a search result could be produced. Also the average number of results per node increased when neighboring nodes were considered. For all 13,503 nodes, the average number of results per node, without neighbors was 0.015. It increased to 0.32 and 8.68 documents when taking 1st and 2nd degree neighbors into account. Adjusting this by only considering the share of nodes producing results the average numbers of resulting documents per node are as follows: The average number of resulting documents per node, without neighbors was 1.2. It increased to 2 documents per node, when taking the 1st, and 9.8 documents when taking the 2nd degree neighbors into account.

A conclusion drawn from these results is, that taking neighbors into account generally increases the share of nodes producing a search output, and also increases the yield of results per node. Since the larger yield is achieved by increasing the radius of neighbors taking into account, it must be assumed that the specificity of results diminishes. Therefore, a trade-off between a large yield and the specificity of results must be achieved. In general, it cna be inferred that, the smaller the radius, fewer but more specific results can be expected; the larger the search radius, the more results can be expected, which tend to be more unspecific.

This trade-off must however be conditional on the connectivity of the local and global graphs, and the number of documents added to it: In the experimental setup with only 200 documents and a single host the yield of search is almost non-existent when neighbors are not considered, so it is critical to take them into account here. With a growing global graph (more clusters, and more links among them), and an increasing number of documents added to them, the number of search results may grow exponentially when considering neighbors. So a dynamic graph-dependant trade-off mechanism should be developed.

Finally, it should be mentioned, that the presented results are of quantitative nature. How well the search results fit with the user's intention, i.e. the quality of search results, needs to be evaluated separately. For now the conclusion presented in [1] is complied with, which states that the TRC serves as a suitable text representation and allows to assume a significant relevance between document TRCs and query TRCs.

6   Summary and Next Steps

The main result of this study was the demonstration of the feasibility of creating a network of cluster files which serves as an extension to a hyperlinked environment such as the world wide web. Even with a single-host setup a simple web search based on the cluster network was implemented and able to produce results. The results suggest fine-tuning with regard to the search parameters, in order to find the right trade-off between numbers of results and contextual specificity of results. In addition, a basic DNS-like Cluster Lookup System was implemented and provided basic functionality to derive cluster hosts as an alternative to blind crawling for clusters. Yet, it still needs to be setup as a distributed system.

As a next step the WebMap needs to be implemented as a distributed network of clusters where the web and cluster server should be provided by multiple hosts. For this purpose it is planned to conduct two further experiments with four hosts based on the same dataset. In the first, the documents will be split into four sets according to their topic and placed on the four hosts. In the second, four sets of mixed-topic documents will be created and distributed to the host.

Finally, numerous aspects have been identified which should be addressed by further research, e.g.:

- Algorithms for cluster negotiation, i.e., merging and removing duplicate clusters and load balancing across cluster servers.
- Algorithm to create sufficient redundancy and backups in the WebMap to avoid isolation and data loss, e.g., for the cases that hosts fail or exit unannounced.
- In order to accelerate TRC computation, methods to further simplify co-occurrence graphs, such that the graph stays well connected but weak edges are removed evenly.
- Suitable metrics for the right trade-off between the number of results and topic specificity, i.e., the right number and choice of neighbors to be considered when deriving search results from clusters.
- A distributed Cluster Lookup System, e.g., deploying the proposed CLS it in a chord network.

Besides error and exception handling these measures are necessary to migrate from a prototype to a productive implementation of the WebMap.

References

1. Kubek, M., Unger, H.: Centroid terms as text representatives. In: Proceedings of the 2016 ACM Symposium on Document Engineering, pp. 99–102. ACM, New York (2016). https://doi.org/10.1145/2960811.2967150
2. Simcharoen, S., Unger, H.: The brain: webengine version 2.0. In: Unger, H., Kubek, M. (eds.) The Autonomous Web. Studies in Big Data, vol. 101, pp. 51–68. Springer, Cham (2022). https://doi.org/10.1007/978-3-030-90936-9_4
3. Roßrucker, G.P., Unger, H.: Webmap: a concept for webengine version 3.0. In: Unger, H., Kubek, M. (eds.) The Autonomous Web. Studies in Big Data, vol. 101, pp. 69–78. Springer, Cham (2022). https://doi.org/10.1007/978-3-030-90936-9_5
4. Stoica, I., Morris, R.T., Karger, D.R., Kaashoek, M.F., Balakrishnan, H.: Chord: a scalable peer-to-peer lookup service for internet applications. In: Cruz, R.L., Varghese, G. (eds.) Proceedings of the ACM SIGCOMM 2001 Conference on Applications, Technologies, Architectures, and Protocols for Computer Communication, August 27–31, 2001, San Diego, CA, USA, pp. 149–160. ACM, New York (2001). https://doi.org/10.1145/383059.383071
5. Srividhya, V., Anitha, R.: Evaluating preprocessing techniques in text categorization. Int. J. Comput. Sci. Appl. **47**(11), 49–51 (2010). http://sinhgad.edu/ijcsa-2012/pdfpapers/1_11.pdf. Accessed 2023-04-17
6. Vijayarani, S., Ilamathi, J., Nithya: Preprocessing techniques for text mining - an overview. Int. J. Comput. Sci. Commun. Netw. **5**(1), 7–16 (2015). https://www.researchgate.net/publication/339529230_Preprocessing_Techniques_for_Text_Mining_-_AnOverview. Accessed 2023-04-17
7. Dice, L.R.: Measures of the amount of ecologic association between species. Ecology **26**(3), 297–302 (1945). https://doi.org/10.2307/1932409. Ecological Society of America
8. Kubek, M., Böhme, T., Unger, H.: Spreading activation: a fast calculation method for text centroids. In: Ben-Othman, J., Gang, F., Liu, J., Arai, M. (eds.) Proceedings of the 3rd International Conference on Communication and Information Processing, ICCIP 2017, Tokyo, Japan, November 24–26, 2017, pp. 24–27. ACM (2017). https://doi.org/10.1145/3162957.3163014
9. Kubek, M., Unger, H.: Towards a librarian of the web. In: Ma, M., Ben-Othman, J., Gang, F., Ooki, M., Cho, G. (eds.) Proceedings of the 2nd International Conference on Communication and Information Processing, ICCIP 2016, Singapore, November 26–29, 2016, pp. 70–78. ACM (2016). https://doi.org/10.1145/3018009.3018031

Authentication in P2P Environment Based on Multi Dimensional Administration Graph

Fariborz Nassermostofi[✉]

Communication Networks, FernUniversität in Hagen,
Unistr. 1, 58084 Hagen, Germany
fariborz@nassermostofi.com

Abstract. Secure authentication in P2P environments has been a central point of attention, since centralized legal infrastructure don't exist in those environments. Different works has been introduced providing required facilities to make such a process possible in a secure manner. Most of these concepts propose the use of Public/private encryption mechanism as a secure authentication approach. Public key cryptography is an established method, which enables secure communication between two parties in insecure environments. Any facility in P2P environment, which aims to use the Public key approach for a secured authentication and communication, must provide a substitute for the required appropriate services in P2P environment, as there is no certificate authority there. This work is concentrated on this matter and provides a concept for an administration graph maintaining a structure for secure disseminating Public key of peers in the whole network. In addition to this structure, required functionalities are provided to perform secured authentication between unknown and known peers in P2P environments using a multi criteria approach. The designed administration graph supports also creation of community and groups with hierarchical permissions for members.

Keywords: P2P systems · P2P infrastructure · Unstructured P2P systems · Authentication · Search mechanism · Lookup mechanism

1 Introduction

P2P systems have commonly known features which are not suitable for secure authentication of a communication adversary in a session. Peers are anonymous and there is no login process. Nevertheless there are a lot of application types, which could make use the strengths of these systems, if there would be a possible secure environment.

In client/server systems only, known and registered members are trusted and allowed to gain access to the network resources. In P2P systems, peers are unknown. In client/server environment, all clients will contact a dedicated server, which can provide certifications issued by a **CA** (Center of Authorization). Peers in a P2P environment communicate with each other without any server and they don't have any certification issued by a CA.

© The Author(s), under exclusive license to Springer Nature Switzerland AG 2023
H. Unger and M. Schaible (Eds.): Real-Time 2022, LNNS 674, pp. 185–212, 2023.
https://doi.org/10.1007/978-3-031-32700-1_18

On the other side P2P systems are scalable, easy to set up and represent the natural way of social live. These systems are also robust and stable by missing some peers. Setting up a client/server with support for many clients will need an organization with policies and required hardware infrastructure. P2P systems don't need any such organization and could be set up without any huge effort for hardware infrastructure.

P2P systems are a point of risk for applications, which need a secure environment. A Secure environment covers authentication of participants and needs mechanisms and facilities to establish trustful communication between participants. All of this circumstances are hard to achieve due to the anonymous nature of these systems and also due to the lack of a CA. What needed, is an environment, in which applications could feel confident about security related matters while fulfilling their business requirements. Overcoming this problem needs a mechanism, which enables peers to authenticate each other in both cases first time (session), when they meet, and in following sessions.

The main problem with authentication in a P2P environments is, that there is no centralized third party instance such as (CA) to attest the authenticity of public key. In centralized environments, system rely on a central third party service, which register the authentication credentials belonging to a certain person or organization. This third party issues a digital certificate after registering authentication credentials of certain person or organization. It provides also the whole set of roles, policies and procedure, which are needed to manage the processes such as create, manage, distribute, use, store, and revoke digital certificates and manage public-key encryption. The required environment used to provide the above facility is called Public key infrastructure referred as **PKI**.

A PKI is a system to provide services to create, store, distribute, use and revoke of digital signatures, which are used to verify that a public key belongs to a certain entity such as a person or an organization. This system consists of a set of roles, policies and procedures to manage those services. According to [1] a PKI consists of:

- Certification policy (CP).
- Certificate Practice Statement (CPS).
- A certificate authority (CA).
- A registration Authority (RA).
- Certificate Distribution Systems (CDS).
- PKI application.

As we know there are some global features in P2P systems, First peers are anonymous. There is no relation between the known ID of peers and the real identification of peers owner. Secondly, in real P2P environments, there is no server for administration purposes like registering peers with valid authentication credentials. This is also the reason, why there is no certification, which peers can use to prove their authenticity. Following this second feature, the third property arises, which denotes, that there is no Login and Logout process as there is no registration. peers join and leave the network occasionally.

The above mentioned features of these type of systems are enough to restrict applications in these area to simple file sharing systems. It is clear that authentication in these systems will have another meaning as in other environments with centralized administration facilities. The process of authentication in P2P environments without any centralized instance will be defined as authenticating a peer repetitive as the one it has claimed to be at the beginning. This is an essential difference between the authentication in P2P environments and in other systems.

Special Features of P2P environments constrain the authentication mechanism to be restricted to use of shared information in this process. One of the most used mechanism here is the use of cryptography to deliver a shared information. Cryptography can be used to establish secure communication between two parties by encoding and decoding messages with keys shared between two parties. The use of public and private key in this area is one of the most established methods. But normally this mechanism needs a centralized facility to provide public key to all other people, who can use it to encode and decode messages exchanged with the owner of that public key. But the centralized infrastructure is not present in P2P systems.

1.1 Motivation

The described Situation in P2P environments regarded to the matter of authentication is motivation of my research subject to find a way making authentication in P2P systems as secure as possible. My aim in this area is to design an approach, in which P2P networking systems would be able to provide required facilities in a decentralized and distributed manner to peers as infrastructure, which can use it to perform a secure authentication.

1.2 Contribution

My contribution in this area is the design of a multi criteria authentication mechanism in P2P systems based on a fully decentralized and distributed public key dissemination system. The proposed mechanism supports a two level authentication, which makes use of the strength of Public/private key approach as the first criterion in the authentication process. As the second criterion, a relation and session based shared information will be used, which is only known by the two participants in a two party communication. The decentralized distribution of public key in this approach will be supported by an administration layer, which is totally distributed over the whole network without any central instance. The structure of this administration layer will be described in Sect. 4.1. The proposed authentication mechanism will be then discussed in Sect. 4.3.

2   Related Works

Some of the most important authentication mechanisms in p2p environments will be shortly elaborated here. These mechanisms are compared finally in the Table 1 at the end of this section. As it will be shown in the Table 1, these mechanisms are observed and compared together using the criteria about centrlization, distribution and multi criteria ability. These criteria will be then explained in Sect. 3.1 in more detail.

2.1   Authentication Mechanisms in P2P Environments

As it is mentioned in Sect. 1, most authentication systems in P2P environments are based on the usage of Public/private key approach. The (k, n) threshold scheme uses public/private keys and divides the secret key into n part. This scheme was introduced by Shamir in [2]. To reconstruct the secret item at least k parts of n is needed. The set of n nodes are also called access structure of the scheme and the system needs a centralized node called dealer to build the access structure [3].

In 1984 Shamir hast introduced in [4] the IBC (identity based cryptography) approach. It uses the users identity as its Public key. A KGC (Key generator node) then create the Private key for the user. The Problem of this method is, that KGC is a centralized node. To overcome this problem Bone and Franklin proposed in [5] another method using more than one Key issuing authorities (PKGs) Nodes. Here they used n PKG nodes to issue the private key for the user. The User then needs to get all parts of its private key from all those PKGs. This method is not centralized but is also not distributed. Additionally the system is slow because the user has to be registered on each PKG node.

Jagadale and Parvat have proposed another optimized approach for IBC in [6]. They combined the strength of (K, n) threshold scheme with IBC approach. So they use a KGC to generate the IDA (Id for the peer A) and also a ProofA. This proofA is a message denoting registration of that peer. They use shamir's (k, n) threshold secret sharing to divide the secret (IDA and ProofA) in to n parts and distribute it among n nodes. The user then needs to obtain the secret from at least k node to reconstruct the ID and proof message. This method is an optimization of other IBC approaches, it is still not distributed in P2P network. In addition within this concept, there is one KGC node to generate the secrets, hence the solution is centralized.

Another mechanism, which is also based on threshold system, is Distributing Security-Mediated PKI (SEMs) and is used by [7] in which each user has a public key (nu, eu) and a private key "du", where n is the product of two large primes. They split the private key "du" in two parts, parts of a user's secret key are dsem, u and duser, u. The part dsem, u is then held by the SEM. All private key operations require the participation of both the user and the SEM.

Other use Certificate Chains [8] technique to distribute public key certificates, which builds a certification graph for the network. Whenever two nodes want to authenticate each other, sub-graphs are merged with or without helper nodes in an attempt to create a certificate chain. In another approach the authors of [9] propose a group based mechanism called troups. In their method groups are built to establish trust relationship between peers. Here a central server is needed to compute those RSA Based credentials. To provide their credentials, they use Zero knowledge protocol between members of troups. They misuse the term trust instead of authentication and it is also not clear trust on what. Peers prove only membership in a group to each other.

The authors in [10] use authentication credentials created during the first registration phase to authenticate each node. They use the public key as the ID of each peer and distribute it via DHT. Hash-based distributed authentication (HDAM) is another approach, proposed in [11] in which mutual authentication is achieved by using web of trust and DHT as base for distributing the Public key. In this method public key of a desired peer will be asked by other peers, which knows it in a chain. The P2P Anonymous authentication (PPAA) is proposed in [12] where peers are familiar and use pseudonymous to one another, when they had met each other once but are otherwise anonymous and unlinkable across different peers. Entities involved in PPAA are the Group Manager (GM) and a set of peer users, or simply peers.

In a PGP system, nodes can get a new valid public key from a trusted node which will be used for communication with that node. However, it is difficult to accumulate all public keys, because PGP does not have a routing map for getting public keys [13]. Authentication method PGP enables a decentralized authentication by using Web of Trust, which is a trusting relationship between nodes. Web of Trust is also used as base for realizing another distributed authentication mechanism in CodiP2P platform. In [14] they build a two level overlay network in which the first level consists of a set of peers connected to a peer as manager. Managers are connected together building another P2P overlay network. The required public key repository is distributed then exploiting the topological capability of their platform.

The Self organized public key management introduced in [15] is another approach and enables an authentication without any centralized service in an ad-hoc network. In a self-organized public key management system, all nodes automatically get new public keys from trusted neighbor nodes in an ad hoc network.

In 2008 an author named Satoshi Nakamoto has published a paper introducing the blockchain approach in [16]. He introduced an electronic payment system in which blocks of information, a nonce, which is a randomly created number, and a block header hash generated by that nonce. The data in the block is signed and tied to the nonce and the header hash. Extracting the block data means finding the number which generates the header hash. This process is then called mining. The longer the chain, the harder is this mining process. The most important aspect of blockchain is, that it is very difficult to change the information stored in Data block [17]. The header hash can be considered as a finger print for the data in block. The problem with approach is, the longer the blockchain gets, it takes longer to decrypt all the information stored in the chains.

Inspired by this method, many other authors have used this approach for authentication purposes in the P2P environment. Hammi et al. have introduced a blockchain based authentication mechanism in [18] to be used in IoT in WSNs (Wireless Sensor Networks). They use Cluster trust and make a node trustworthy by all clusters, if the node is authenticated in one cluster. Using blockchain they provide the information stored in the block is available to all participating nodes.

Liu et al. [19] have introduced a new G2G authentication scheme based on physical unclonable functions (PUFs) and blockchain. They use the fingerprint of chips to create a secret key for a node and divides it into m shards of equal size and store it together with the corresponding authentication data on them. The authentication consists of Challenge and response part of PUFs and the node will then create m transfer transactions to m nodes. The chain is created by sending the shards to each node of the m selected nodes. This is called the authentication information and is dispersed between m nodes in chain of blocks and can be considered as authentic when decrypted.

Mukhandi et al. propose in [20] a Device identity Management with consensus authentication for IoT Devices based on Blockchain approach. In their proposed method, they create a digital ID for each device and encrypt it using the Private key of each device to. Additionally each IoT device stores a Merkle tree of the public keys of all other nodes and the digital ids of all blockchain nodes. Every new node in the network gets also the same information sent to store it. The digital identity used here consists of device name, firmware, MAC address and configuration files and a timestamp. This created identity is then stored in the blockchain ledger. To authenticate a requester node sends an authentication request, its request will be broadcast to all IoT nodes in the chain to verify its identity and so achieve a consensus agreement about the identity of the requester node.

In Table 1 some of the most important authentication methods in P2P environments are listed. The above elaborated mechanisms are considered regarded to three criteria as points of view in the present work. The first criterion is whether a method is using a centralized server or approach. The second criterion is whether a method is distributed in a way, that authentication credentials are really dispersed in the environment and not stored on a single instance. The third criterion is whether the mentioned method is using a multi criteria approach or just one criterion approach in the process of authentication.

With respect to the criteria declared above, we see that some of the mentioned mechanisms are using a kind of centralized mechanism and some other not. In most works they use the term decentralized and distributed in an interchangeable manner. So when they use the term distributed, they really are talking about a decentralized approach. But a decentralized approach is not imperatively also distributed. As an example Distributed PK via DHT is decentralized because they are using a DHT (Distributed hash table). In this mechanism the DHT is distributed and nothing else. The atomic and particular information, in our case

Table 1. Authentication Approaches in P2P environments

Authentication mechanisms	Centralized	Distributed	Multi criteria
Threshold Cryptography	Yes	partial	No
IBC (Identity-based cryptography)	Yes	No	No
IBC with n PKG Nodes	No	partial	No
IBC with Threshold Cryptography	Yes	partial	No
Distributing Security Mediated PKI	No	partial	No
Certificate Chains	No	No	No
Trust Model based on Troups	Yes	partial	No
Distributed PK via DHT	No	No	No
Distributed PK (HDAM)	No	No	No
P2P Anonymous authentication (PPAA)	Yes	No	No
PGP Authentication model	No	No	No
Distributed authentication CodiP2P	Yes	partial	No
Self organized public key management	No	partial	No
BCTrust	No	partial	No
G2G authentication	No	partial	No
Blockchain based identity Management	No	yes	No

the authentication credentials like Public key, are stored on one peer. Which is then still centralized and absolutely not distributed. Only the way how to find a server holding that public key is managed in a distributed manner. So having this point of view in the observation above in mind, we see that almost all of the mentioned methods above, except of Blockchain based identity Management, are not using a decentralized and distributed authentication mechanism. In Addition, as it is also marked, n one of the methods above is using a multi criteria approach.

2.2 Anonymous Authentication

As it is described in Sect. 1, one of the features of P2P networking system is the anonymous nature of peers behavior. There is no registration and Login process in such a networking systems. The Message exchange performs in an environment, in which normally sender and receiver are anonymous. This is called anonymous broadcast communication. The problem within this type of the communication is clearly the matter of authentication.

The key-point here is, that a sender can communicate with a receiver, without the receiver learning the identity of the sender. There are a few successful system in this area like, Tor, Tarzan, Freenet, and FreeHaven [21].

This area is related the aim of this paper, where two peers which are anonymous and can not deliver any secure identification credentials, due to the lack of a server and a CA, need to authenticate each other. The problem of anonymous broadcasting communication is intensively elaborated and studied in Dining cryptographers problem (DC-nets) [21], which will be discussed in next subsection.

Dining Cryptographers Problem. Dining cryptographers Problem or shortly DC-nets protocol is first introduced by David Chaum in his work in [22]. He illustrated the introduced algorithm with the famous example of the problem, that few friends were going for a dinner and at the end they wanted to know who has paid the bill for that dinner; one of them or NSA as example. If one of them has paid, they didn't want to know who. Within the proposed protocol by Chaum, every two person share a secret key. The first person votes then with 1 or 0 as her statement whether she has paid for the dinner or not and The result will be sent to the second person encrypted using the shared secret key.

The second person will also vote and performs a Boolean XOR operation on her vote and the vote of the first person. The result determines whether one of them has paid or not. when the result is 1, one of them has voted with 1 and has paid for the dinner. Otherwise no one of the two person has paid for the dinner. The result of this two person will be sent then to the next one in the round and so on. At the end of process, the result is 1 or 0 and they will be able to know whether one of them has paid for the dinner or not without knowing who has paid in case that the final result is 1.

The secret key will be created between every two communication adversary using Diffie-Hellman protocol [23], which ensures the creation of a secret information between two parties without sending it through the communication media, whatever it might be. Within this protocol both parties create a secret key, under the combination of a shared key, known to both of them, and their own private key. At the end of the protocol, both parties are in possession of the same secret key, which is known only by the two parties.

As it is described in [24], DC-nets are normally used for sender-anonymous, point-to-point communication. Especially when there is more than two communication partner. Message packets will be encrypted using the same secret key on both side.

Due to the nature of this protocol and the need to establish a private communication between every two peers, The protocol needs a higher effort $\Omega(N\,2)$, both computational and communication in large groups [25], which is the reason why this protocol is not used widely.

Zero-Knowledge Protocol. As a tool in the area of cryptography, Zero-knowledge protocol plays an important role, where not the desired shared information, but the proof of the possession of that information will be delivered. This protocol was first proposed by Goldwasser et al. in [26]. This is an interactive proof system (P,V), where P is a prover and V is a verifier. A zero-knowledge proof must satisfy three properties [26].

1. Completeness: that it should be possible to "prove" a true theorem

2. Soundness: that it should not be possible to "prove" a false theorem.
3. Zero-knowledge: that is by proofing the theorem, no information about the theorem itself is revealed.

An important factor in a Zero-knowledge protocol is the choice of the mathematical problem as the basis of the protocol. One of the most used mathematical problem is the graph isomorphism problem and hamiltonian cycle graph. The interactive process of proofs between the prover and verifier using the graph isomorphism and hamiltonian cycle graph proceeds normally with the creation of a graph G_0 by the prover The prover knows the hamiltonian cycle graph to graph G_0. Now in each interaction the prover needs to create another distinct isomorph graph of G_0 and show as an example that she knows the hamiltonian cycle of the created isomorph graph, which is then the proof that she knows the hamiltonian cycle of the original graph.

Due to the survey in [27] The protocol is used in various forms, such as Zero Knowledge Password Authentication Protocol or a more sophisticated form like Zero knowledge Password Authentication Protocol with Public key encryption. The protocol is even used in P2P environments for authentication as it is used in [28] to prove the possession of the right authentication credentials. They use a two phase scheme in which in the 1. phase a registration of peer at a management site will be performed. In the second phase peers authenticate themselves without the need of the management peer using Zero-Knowledge protocol. The draw back of their solution is the need for a centralized management peer.

While the ZKP model is an interactive protocol in which messages will be sent to the partner several times, another non-interactive variant of this protocol is proposed in [29]. In their solution all required information of every interaction will be send in one message consisting of different segments. Every segment consists of one representation of an isomorph graph to the original graph. While the first segment is not encrypted, all other segments are encrypted each with different key. The encryption key of each segment depends on the previous segment. Every user who wants to decrypt the last segment, must decrypt all other segments. Both party agree to use a one-way hash function to define the challenge that a receiver must solve on each isomorphic graph. Additionally the same hash function will be used to define the encryption key for each message segment.

As it is introduced in [29] operations, the prover has to perform to verify the correctness of proof, are defined as followed:

1. Process the first segment of the message, which is not encrypted.
2. Compute, using the hash function, the challenge that matches the information included in the segment.
3. Check whether the response corresponds to the challenge and the isomorphic graph.
4. From the challenge, compute the key it has to use to decrypt the next segment.
5. Apply Steps 2 through 4 until the last segment, which once deciphered contains the information needed to establish the desired result.

The security level of this protocol depends on the number of segments in the message. The more segments are used in the message the higher is the more complex is to reach the last segment. This model is known as Non-Interactive Zero-knowledge proof (NIZKP).

3   Requirements on the Mechanism

The proposed mechanism is a multi criteria authentication approach using distributed Public key for first contact testified by some other peers for the first contact and use of the public key and a created shared secret between familiar peers on other following sessions. There are some requirements for the proposed concept to fulfill the desired authentication approach. These requirements will then determine the set of operations the system must provide, which will be introduced and discussed here.

3.1   Conditions

Performing secure authentication in applications based on P2P environment needs a system fulfilling following conditions.

1. Due to the strength of **Public/Private key encryption**, this approach should be used for both authentication and communication processes to make them secure.
2. In absence of CA and regardless of the used infrastructure for the P2P environment, peer's public key must be **disseminated** and **distributed** in the system.
3. Due to some features of P2P networking environments like anonymity and lack of any centralized registering services, it is clear that a simple one criterion authentication is not secure enough, so a **multi criteria authentication** is desired to ensure the process.
4. Peers in the system must be able to perform **repetitively secure authentication**.

The dissemination of public key will be achieved using a distributed data structure. This structure will be explained in Sect. 4.1. There are some condition related to this structure.

1. The structure must be **accessible** overall in the network.
2. To ensure the **authenticity** of public keys, only their owner must be able to manipulate them.
3. The structure has to be **reliable**. It must be available even if one of the some member of the structure are not present at a certain time.
4. The required structure needs to be **scalable** and grow with the size of network.

3.2   Operations

The set of all operations needed to fulfill the above condition on the proposed approach is also divided in two categories. The first category is related to the operations needed to build up the distributed structure of the approach.

Most important operations in the first category are regarded to building up the administration structure and disseminating the public key in it and manipulating and querying the public key in it.

To build up the distributed structure an operation for **distribution of PK** is needed, which builds up the required globally accessible distributed structure recursively. The **finger** checks whether a peer belongs to administration layer of a certain peer. The **setRepeat** operation informs a peer, that it has to take the role of replication for another peer in administration graph for a certain peer. This is needed to increase reliability of the distributed structure. This distributed structure needs also to deliver the public key of a certain peer in the case it is asked by any arbitrary other peer, which wants to communicate with that peer. So asking for public key and delivering it is also seen in a **providePK** operation. In case of changing the public key by a peer, the distributed structure must be informed and updated. This will be performed by using the **updatePK** operation.

The second category of operations is related to the matter of authentication. As long as a multi criteria authentication approach is used in this solution, there are also two different type of methods to authenticate a peer regarded to each criteria.

– Authentication for unknown peers (First contact authentication)
– Authentication for familiar peers (Next time authentication)

Authentication for Unknown Peers. The first criterion uses only public key of the targeted peer. In the present solution there are two different authentication situation. The first case is related to the first contact between two peer, when they are connected together for the first time. In this case the only possibility they have to authenticate each other is the public key. In this case peers need to get the surety of authenticity of the communication adversaries public key. To get the surety about that matter, the requester can ask a few other peers to testify the public key's authenticity of a certain target adversary. This request will happen using the method **confirmPK** by asking at least two different peers. One could be the root peer's administration graph of the source peer and the other one could be the root of target peer's administration graph as long as they are different. Otherwise any other peer in the network could be asked as the second peer.

Authentication for Familiar Peers. Once the authenticity of a public key by it's owner is proved, peers can start communicating together using their public keys. They are now familiar with each other and can create shared secret between themselves and use as a second criterion in each next session.

4   Proposed Approach

In this section an authentication approach will be proposed, which consists of a two level authentication. The first level is the process of authentication using public/private key procedure. This procedure is used both for first authentication and as the first criterion and in each re-authentication process. The second level uses a relation and session based shared information which is only known to both communication partners.

The first level of the authentication rely on public key procedure. Recalling from the defined condition on the concept in Sect. 3.1, the public key must be fully distributed. To disseminate and distribute this information (public key of peers) over the whole system an administration layer is proposed, which will be described in Sect. 4.1. After the description of the admin layer, the process of multi criteria authentication will be described in Sect. 4.3.

4.1   Definition of Administration Layer

To disseminate the public key in a decentralized and fully distributed manner, an administration layer is designed. The admin layer is an acyclic directed graph of connected peers, which contains certain information about a peer to make other peers able to find the public key of that peer in order to be able to perform an encrypted communication with it. This structure includes practically all peers in the environment in the maintenance of administrative infrastructure.

Given N peers in the P2P networking space there will be N administration graphs, each providing all required information about the structure of a peer's administration graph. These information are stored in a table on each peer within this graph. Information stored on each node in an admin layer graph are (Table 2):

Nodes involved in the administration graph have each one predecessor and some successors. The administration graph for each peer is constructed during its

Table 2. Information stored on each peer in Admin layer

Item	Description
ID	ID of the peer which is the owner of the public key
Public key	Public key of the peer
Resources[]	Resources the peer provides
Groups[]	which is a semicolon separated list of the set of a groups, a peer is member of, together with its role in that group separated by colon.
	The name of the groups must be fully qualified in order to provide groups hierarchy information and group's pmask.
	Example (group1.group2:member:1,group1.group2.group3:admin2)
Rootpeer[]	Pointer on root peer in the administration graph and its repeater
Predecessor[]	Pointer to the previous peer in the chain and its repeater
Successors[]	Pointers to next peers in the chain and their repeater
Repeater[]	Pointers to peers acting as redundant for each admin peer

first join to the network. The structure of an administration graph is determined by the following definition:

P is the set of peers in the P2P networking system

G is the administration layer of the networking system

g is an administration layer graph

V set of authority nodes in an admin layer graph

E is the connection pointers between nodes in an admin layer graph

D is the distance parameter set during the bootstrap phase

Given a peer $p_i \in \mathbf{P}$, there will be the following relation.

$$\forall\, p_i \in \mathbf{P}\; \exists\, \mathbf{g}_i\{\mathbf{V},\mathbf{E}\} \in \mathbf{G} \vdash \{\mathbf{V}\} \subset P \;\wedge$$
$$\mathbf{maxlength(E)} \;\leq\; \mathbf{D}$$

And following relation is deduced from the a above definition:

$$g_i\,\{\mathbf{V},\mathbf{E}\} \;\neq\; g_{i+1}\{\mathbf{V},\mathbf{E}\}$$

From the equation above we understand that for each peer in the P2P system there will be one administration graph within the P2P system, where the distance between all its nodes $\leq$ the global parameter Distance is. Nodes in each administration graph may also be member of other administration graphs but generally the administration graph of one peer is not equal to the administration graph of another peer.

Communities. Using this structure the entire P2P environment would be connected. Here there is also a possibility to create an overlay networking environment using the **community** facility in it. Any Peer can define and start a certain network layer, such as School, university or any kind of organization. A Community is created by defining a new group by any arbitrary peer. Each peer is able to set other peers as member of its group. Groups can be then defined hierarchically and build up the entire organization.

Peers are member of a group inside a community and can themselves create a new group under the group they are member of. Groups need to have pmask (peer-resource-access mode mask)specified to define the access level of members within the community. Each provided resource by a peer in the community is related to a group and will be assigned a certain pmask in order to define access level of other peers to that resource. The pmask used in this concept defines read and write access for different kind of peers analog to pmask of files in Unix systems. In this context peers will assigned one of the three following roles related to a group:

– Owner and admins of a group

- Member of a group
- Other

The defined pmask will not provide the possibility of execution of a resource. It only describes a read and write access for each of the above recognized items. The access level is upward and a resource with a certain pmask has already lower pmask in it. The pmask of a resource is defined as followed (Table 3):

Table 3. pmask of groups

Owner	admin		Group		Other	
1	2	3	4	5	6	7

The meaning of this pmasks are as follows:

- **1** The group's owner has always read and write permission
- **2** The group's admin have read permission
- **3** The group's admin have write permission
- **4** The group has read permission
- **5** The group has write permission
- **6** The others have read permission
- **7** The others have write permission

In Order to be able to use this pmask in a hierarchically constructed environment, member of a subgroup will inherit the pmask settings of the resources provided in a supper group. That means a resource with read permission for the group *3* is also readable for peers which are related to subgroups of resource provider's group. Considering this definition and regarding the content of administration table's record a group definition might look like following statement:

Community.Group[A]:member:4,
Community.Group[A].Group[A1]:admin:2,
Community.Group[A].Group[A2]:admin:3,
Community.Group[B]:owner:1

which denotes that the actual peer is member of group **A** with only read permission and admin of group **A1**, which is a subgroup of group **A** with only read permission but admin of another subgroup of group **A**, group **A2**, with write permission and created a group **B** and is its owner. The hierarchy related to the above group description is shown in Fig. 1.

Peers which are member of group **A1** inherit the pmask of a resource provided in group **A** and have read access to a resource with a pmask **3** provided in group **A1**. Peers which are member of group **B** will not have any access to the above resource unless the group will get the pmask **5** or **6**.

On the other side members of group **A** have no access to resources provided on groups **A1** or **A2** unless these resources will get the pmask **5** or **6**.

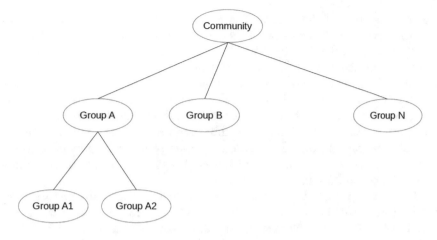

Fig. 1. Community hierarchy

4.2   Construction of Administration Graph

During the bootstrap of the system some peers with long time availability will be dedicated as initial connection peers for P2P system. Each new peer joining the system contacts one of this nodes. These initial peers may accept the connection attempt of the new peer and connect them or forward them to another server (due to decision based on load balance or location properties). In any case the node, which accepts the connection attempt of the new peer, is the entry point of the peer in the environment. This peer is also called the root peer of the administration graph for the new peer in the environment.

Root Peer in the Graph. At its first join after creation of a valid ID (the process of creation an ID in the environment is not covered in this proposal) and pub-lic/private key, the new peer sends this information to the peer it is connected, also its root peer, in order to distribute them. This root peer, stores the infor-mation of the new peer in its local administration table. Each peer in the P2P network maintains a table which contains administration related information for each administration graph, that peer is member of. Content of this tables con-structs for different peers the administration layer graph. The root Peer marks itself as the root of administration graph for the joined peer.

Repeaters for Each Admin Peer. It also send the information of the new peer to maximum 5 of its direct neighbors to be marked as its repeater for the actual peer in its administration graph. These repeaters ensure more robustness of the administration graph if one of the main admin peers is not present at any time, there will be with a higher probability one of its repeater for each of the administrator graphs, it is member of each repeater inserts the exact information for the actual peer in its local table as the admin peer.

Recursive Building of the Graph. The root peer then sends the ID and public key of the new peer as parameter of the method distribute to all peers connected to it using a parallel BFS algorithm. As additional parameters, the parameter *"distance"*, and its own ID will be passed through. The value of this parameter *"distance"* determines after how many steps a new member of administration graph for the new peer is to be added to the graph. After each step the value of parameter "distance" will be decremented by 1. Reaching the value 0 causes the target peer to start a search in its neighborhood for the existence of a member of the administration graph for the newly joined peer. If there is no member of administration graph in a radius *"distance"*, then it will insert the information of the new peer in its administration table.

When a peer accepts to be member of administration graph of another peer, it has to insert several information in its local administration table. First to fulfill the main purpose of the administration graph, distributing the peers public key, it will contact the new peer and asks for confirmation of the public key. This confirmation ensures that only the owner of the public key is able to start a distribute operation and set or update its public key and no one else. Additional information are the ID of the root peer of the actual peer and its repeater, the id of the previous peer in the administration graph and its repeater.

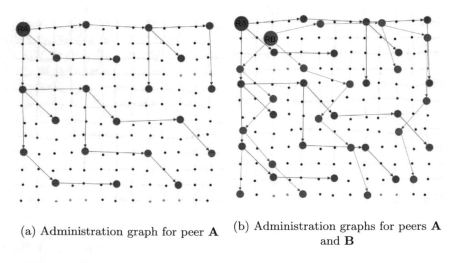

(a) Administration graph for peer **A**   (b) Administration graphs for peers **A** and **B**

Fig. 2. P2P environment with Administration graphs for 2 peers **A** and **B**

Figure 2 illustrates 2 administration graphs in a simple p2p networking environment. Black nodes denotes peers in the network and the grid indicates connections between node members. For the sake of simplicity and visibility, a simple 2 dimensional environment is chosen to illustrate the construction of an Administration graphs. In figure a only one administration graph for a peer **A** has been created with a Distance parameter **H = 3**. The constructed administration graph for a next peer **B** is then shown in figure b. For both peers we see that each administration graph has an initiate root peer, which then starts the construction of the administration graph for each of those 2 peers recursively.

As it is shown here the administration graph for each of those 2 peers is completely distributed over the whole network in an decentralized manner. Every peer searching for a resource provided by one of these two peers will fined definitely in a radius of **H** peers a peer, which is member of the administration graph of those 2 peers and will be informed about the provided resource by them. This way every peer will find in a radius **H** the information about all resources provided by all peers in the network. This means a global view in a completely unstructured P2P environment.

Predecessors a Successors. Every member in the administration graph for a peer has to maintain a list of its predecessor peer and its successors together with their repeater that means for a predecessor we will have an entry in the following form:

predecessor:repeater1,repeater2,repeater3,repeater4,repeater5;

And the entry for successors would be in the following form:

successor[1]:repeater1,repeater2,repeater3,repeater4,repeater5;
...
successor[n]:repeater1,repeater2,repeater3,repeater4,repeater5;

Each peer then determines also on the same way like the root peer, its repeater and starts the distribute method similar to root peer again recursively. This algorithm continues until overall in the P2P networking system at least one peer as administration graph for each peer is given. If the admin peer gets some successor during the recursive call of the method distribute, it will put their ID also in the local table as described above. Each successor delivers the ID of its repeater, if exists to its predecessor. The list of successors of an admin will also be passed through to the repeater of that administration node. The set of all administration graph composes the admin layer of the P2P environment in order to provide in a fully decentralized and distributed manner all for a successful authentication required information to the whole network.

Protocol. Before the algorithm of methods can be explained, some idioms used in the algorithms need to be explained. These idioms are listed and explained in the Table 4.

The functionality required to build up this admin layer consists of a few methods. These methods are designed to fulfill together with the ability of the designed administration graph the requirements on such a system as it was declared in Sect. 1.2. Table 5 shows a list of these methods with a short description on their operation. Some of these methods, consisting the main algorithms are then explained.

Table 4. Idioms used in Admin layer protocol

Idiom	Description
ID_Root	ID of the root of an administration graph
ID_Caller[]	ID of the peer which has invoked the method and its repeater
ID_Peer	ID of the newly joined peer
ID_Repeater[]	ID of the Repeater peers for the actual admin peer
ID_predecessor[]	ID's of the predecessors of the actual peer in administration graph
ID_Succ[]	ID's of the successors of the actual peer in administration graph
PK_Peer	Public key of the newly joined peer
Distance	The actual distance from the last admin in administration graph

Table 5. Methods used in Admin layer protocol

Method	Description
distribute	distributes recursively the public key of a Peer in the network
providePK	delivers the public key of a certain peer on behalf of any peer
updatePK	updates the administration graph with the new public key of a peer
validatePK	asking the owner of the public key for validation of the given key
confirmPK	asking at least 3 peers to confirm the Public key of a peer
setRepeat	Marks direct peers as a repeater in an administration graph
finger	looks for admin peer within MAXDISTANCE radius
insertGraph	inserts the information of an administration graph in local table
updateGraph	updates the administration graph
updateRepeater	updates Repeaters with Successors

The Method finger will look in Distance defined in **MAXDISTANCE** for the existence of an administrator for the given ID. The algorithm for this method is as follows:

```
===================================================================
Method finger(Distance, ID_Peer)
===================================================================
Start
  IF (isAdminMember(ID_Peer)) // is the actual peer an
                              //  admin of the new peer
    RETURN(Actual ID)
  ELSE
    IF (Distance == MaxDistance) // is maximum radius reached
      RETURN(NULL)
    ELSE
      Distance--
      //looking further for an admin
      FOREACH(ID of direct neighbors[])
```

```
        ID_admin = finger(Distance, ID_Peer)
        IF (ID_admin != null)
          Break
        END IF
      END FOREACH
      // returning either null or an ID of an admin
      RETURN(ID_admin)
    END IF
  END IF
END
```

The setRepeat method informs direct neighbors of a peer as a repeater for its role as admin in administration graph for a certain peer.

```
==================================================================
Method setRepeat(IP_Root, ID_Caller, ID_Peer, PK_Peer)
==================================================================
START
  i=0
    FOREACH(MAX(5) of ID of direct neighbors[])
      ID_return[i++]=insertGraph(ID_Peer,PK_Peer,ID_Root,
                                ID_caller[])
    END FOREACH
  return(ID_return[])
END
```

The updateRepeater method in Opposite to the setRepeat does not need to select 5 neighbors randomly. It knows which are the repeater and has only to update the successors of the appropriate admin peer on its administration table. This function is described below:

```
==================================================================
Method updateRepeater(id_Repeater[], admin_ID, ID_Peer, ID_Succ)
==================================================================
START
  i=0
    FOREACH(id_Repeater[])
      ID_return[i++]=updateGraph(admin_ID, ID_Peer, ID_Succ[])
    END FOREACH
  return(ID_return[])
END
```

Distribute method is one of the most important functionalities of the admin layer and is responsible for building up the desired administration graph for every peer. The algorithm of this method is described below:

```
==================================================================
Method distribute(ID_Root,ID_Caller[],ID_Peer,PK_Peer,Distance)
==================================================================
START
```

```
IF (ID_Root == own_ID)
  id_Repeater[] = setRepeat(IP_Root, null, ID_Peer, PK_Peer)
  ID_predecessor[] = null
  // predecessor and its repeater
  ID_CALLER[] = own_ID + ':' + id_Repeater[]
  Distance = 0
ELSE IF ( (Distance == MaxDistance+1) AND
          (!finger(Distance-1, ID_Peer)) )
  // validating the Public key by asking the owner
  IF (validatePK())
    ID_predecessor = ID_Caller
    id_Repeater[]=setRepeat(IP_Root,ID_Caller,ID_Peer,PK_Peer)
    // predecessor and its repeater
    ID_Caller[] = own_ID + ':' + id_Repeater[]
    Distance = 0
  ELSE
    RETURN(0) // the public key could not be validated
  END IF
END IF
Distance--
i=0
FOREACH(ID of direct neighbors[])
  ID_Succ[i++]=distribute(IP_Root,ID_Caller[],ID_Peer,
                          PK_Peer,Distance)
END FOREACH
// either root or an admin peer
IF ( (root_Peer) OR (admin peer) )
  insertGraph(ID_Peer,PK_Peer,ID_Root,ID_predecessor[],
              ID_Succ[])
  updateSucc(id_Repeater[], own_ID, ID_Peer, ID_Succ[])
  return(own_ID + ':' + id_Repeater[])
ELSE
  return(ID_Succ[])
END IF
END
```

Validation of the Protocol. The above listed algorithms shows some of the must important methods in Admin layer concept which ensure fulfilling the requirements mentioned in Sect. 3.1 regarding the functionalities of Administration graphs. The distribute method ensures the fully distribution of the public key of a new peers in the environment in a way that every peer can find a public key for any arbitrary peer in a radius **MAXDISTANCE** hops within the environment around it. This method ensures the fulfillment of two of the main conditions namely distribution and accessibility.

By invoking the method validatePK by an admin-candidate it can get the surety that the public key given in the parameters of the method distribute belongs really to the owner of that public key and is not distributed by a malicious peer, so the security of the admin layer is guaranteed.

The reliability of the admin layer is achieved by replication of an admin peer within the administration graph so that for each admin peer there will be 5 replications. The list of the replicated admin peers is also published in the administration graph so, that when one of the admin layers are not present in the network at any time, one of the repeated admins can response to public key look up attempts of peers in the network. So the reliability is here guaranteed.

As long as the Administration graph in the admin layer is fully distributed and grows with the grow of the network and is not stored and maintained on single peers with huge need of resources, the structure is scalable and there is no limitation regarding the size of an administration graph. The scalability of the distributed structure is also guaranteed.

Advantage of the Model. The distributed admin layer as it is proposed here has some advantages. The admin layer can be used as a middle ware regardless of the underlaying infrastructure in the P2P both for structured and unstructured environments. Using this middle ware for the authentication, there will be no single point of failure regarding the authentication credentials. Using the proposed protocol functionality the structure is safe against unwanted and malicious manipulation. Using this admin layer, even in case of unstructured underlaying infrastructure, all peers in the system will have access to the authentication credentials of all other peers.

4.3 Multi Criteria Authentication

The Proposed multi criteria authentication mechanism uses two complementary approaches to authenticate a peer as it was introduced first to the P2P environment and to the communicating peer. The process of authentication in this concept consists of two different approaches in different situations. When two peers had no contact at any time before and met(contact) each other for the first time, they will have no shared history and the only authentication method between those peers is the use of provided public key of them. From the moment of a first authentication they will have a shared history and can rely for further authentication on an additional criteria in the process of authentication, which then will be a relation and session based shared information between them.

First Authentication Criterion. Public/private key encryption mechanism will be proposed for use at the first contact and as the first authentication criterion in further sessions. Access to the public key of a peer at the moment of first contact happens through a request in P2P environment. The public key of each peer within the P2P environment is kept in distributed manner on the admin layer as it was described in Sect. 4.1.

With invocation of the method ProvidePK(), an arbitrary peer can ask an admin peer as member of the administration graph of any target peer for its public key. Normally a peer will be listed as the result of a search for certain resources some might share or provide. When a peer wants to contact another peer providing some resources, it need first to have the public key of the target peer in order to be able to send him encrypted message as the first contact attempt. This is important, as the target peer, which should be the owner of the Public key, is the only one which can decrypt the incoming message.

The peer, which listed all resources of the target peer to the requester is one of the admin peers of that target peer. This admin peer can answer to the requester calling the method providePK() with the desired public key. It also can return the ID of target's peer root node in its admin graph as long as this root node is not the root node of the requester, it can be used to verify the authenticity of target's Public key. The third peer which can be used as verifier could be the root of the requester self other wise any randomly selected peer in the network could be used to verify it.

Once the authenticity of the public key is verified, the requester can send a message to the target peer containing contact request. After the first authentication using the public key is performed, both parties will create a relation/session based authentication credential which will be described in next subsection to be used as second authentication criterion during further sessions. So in each new session, the process of authentication consists of using the relation/session based authentication credential using the public key of those peers.

Second Authentication Criterion. The mechanism proposed in this section uses features of the both described mechanisms in cryptography in Sect. 2.2, to ensure a secure authentication between peers, which are known to each other. Inspired by the DC-nets protocol, a shared key will be created between each two communication parties after the first authentication to be used in the next session, using Diffie-Hellman protocol.

When peers establish connection at the time 0 contacting each other and perform a standard authentication by using their Public/private key credentials, from that point they know each other and are familiar. At the end of the first authentication process, both parties can create relation /session based private credential to be used for creation of further additional authentication criterion. These credentials will be used to authenticate both parties to each other. But instead of exchanging these credentials, only the proof about the possession of the right credential will be delivered. So the authentication process between two familiar peers consists of the two following steps.

1. Producing of relation/session based shared secret.
2. Proofing the possession of that shared secret.

The first component is used to perform two tasks in the process of authentication between familiar peers, which are:

1. Creation of one time secret key using Diffie-Hellman protocol.
2. Creation of one time authentication credential.

As it is mentioned in Sect. 2.2 in DC-nets, a secret key will be created between every two peers using Diffie-Hellman protocol. This key is considered to be a one time pad to be used, during the next authentication process between them. So at the end of authentication process of the first session also s_0, the secret key sk_1 has been created and both adversaries hold them. During the next session, when they contact each other, the secret key sk_1 will be used as a one time pad in the authentication process. At the end of the authentication process in every sessions s_n the secret key for use in the next session also sk_{n+1} will be created.

So for every two peers $P(a)$ and peer $P(b)$, there will be a secret key sk_n for use in session s_n such that following relation exists:

let **S** be the set of all authentication sessions and
SK be the set of all secret keys

$$\forall \, \mathbf{s} \, \in \, \mathbf{S} \, \exists \, \mathrm{sk}_n \vdash \, \mathrm{sk}_n \neq \mathrm{sk}_{n+1}$$

The second step is creation of an authentication credential. The credential for session s_n will be created at the result of a boolean XOR Operation between the sk_n and a known information between the two peers. This Information does not need to be secret. one of them can even send a large number as the information to the other peer. The secret key of that session also sk_n will be used to perform the boolean XOR operation with that large number. Using this method there will be following relation:

let **S** be the set of all authentication sessions and
SK be the set of all secret keys and
SAC be the set of all shared authentication credentials and
I be the set of all known shared Information and

$$\forall \, \mathbf{i} \, \in \, \mathbf{I} \, \exists \, \mathbf{i}_n \vdash \, \mathbf{i}_n \neq \mathbf{i}_{n+1}$$

which denotes that the used information to create the authentication credential for each session changes for every new session and:

$$\forall \, \mathbf{s} \, \in \, \mathbf{S} \, \exists \, \mathbf{sac}_n \, = \, \mathbf{sk}_n \bigoplus \mathbf{i}_n \, \& \, \mathbf{sac}_n \, \neq \, \mathbf{sac}_{n+1}$$

From the above relation we deduce, that authentication credential used for every session is changed and is not equal with the previous one. Now when the proofer has been able to create the expected result, she needs to inform the verifier about the correct result, she has been able to produce. The verifier, on the other side, can produce the same result. Both parties are in the possession of the same one time produced shared information, which is called in this schema the authentication credential.

Using the NIZKP, the prover will send only the proof that she is in possession of the right information. As it is explained in Sect. 2.2, the prover will create a graph representing the authentication credential and produce then the n isomorph graphs corresponding to the original graph. The verifier peer creates also the same representing graph of the authentication credential using the same algorithm.

At this point the prover can prepare the n segmented proof message using NIZKP as is explained in Sect. 2.2 and send it to the verifier, which then can verify the received information. The whole schema consists of the following protocol functionality:

1. **create_pk**: which creates the private key hold by each peer.
2. **create_shk**: which creates the shared key initiated by the verifier on both side.
3. **create_sk**: which creates based on the shared key and private key, the secret key.
4. **create_graph**: which creates a graph representing the authentication credential.

The algorithm used to create the secret key for every new session looks like the code here:

```
=====================================================================
Method create_sk()
=====================================================================
START
   pk = create_pk()                \\creating the private key.
   shk = create_shk()              \\creating the shared key.
   send_msg(shk)                   \\sending the shared key to partner.
   sk_1 = boolean_XOR(pk, shk)     \\performing the boolean OR.
   \\operation between the private key and the shared key.
   send_msg(sk_1)                  \\sending the result to the partner.
   sk_2 = receive_msg()            \\receiving the intermediated key
   sk = boolean_XOR(pk,sk_2)       \\creating the secret key
   return(sk)
END
```

The authentication process will then use the following algorithm:

```
=====================================================================
Method authenticate(number n)
=====================================================================
START
   large_number= random() \\creating a random large number
   send_msg(large_number) \\sending the large number to partner
   sac = boolean_XOR(sk(n),large_number)  \\creating the desired result
   \\based on secret key of Session n and the larg_number
   graph = create_graph(sac)  \\creating the representatin graph of sac
   result_p = receive_msg() \\receiving the NIZKP prepared proof message.
   IF (check(graph, result_p)) \\check if the result from partner
      create_sk(n+1)           \\corresponds to the result created localy
   return(1)
   ELSE
      return(0)
   END IF
END
=====================================================================
```

although the exchanged information in our example is a randomly created large number, in different environment any other shared information between both parties can be used, such as the number of communication packets exchanged in the last session.

Validation of Authentication Process. In Sect. 3.1 4 main conditions has been declared on the global authentication system. According to the first condition Public/Private key encryption has been used in this solution for both authentication and communication process.

The second condition determined a distributed approach. The proposed middle ware for a distributed public key guaranties a fully distributed approach, in which the desired information and not the way to find that information is distributed. The validation of operations and protocol of the admin layer as middle ware is given in Sect. 4.2. Using the proposed admin layer, a fully distributed authentication approach is guaranteed.

As long as the process of authentication between two peers in the proposed mechanism makes use of two different criteria, which are Public/private key encryption mechanism as the first criterion and a relation/session based authentication credential as the second criterion, the proposed authentication approach is multi criteria. Only the authentication at the first contact will use a one criterion, which is public/private key encryption approach. To get the surety about the authenticity of that public key three peers will testify it.

The forth condition was the possibility of repetitive authentication in the anonymous environment of a P2P networks. This is criterion is fulfilled again by the usage of multi criteria approach specially the usage of the relation/session based authentication credential created at each session by the peers in a communication. Providing this credential next time will ensure the secure authentication in the next session.

The matter of secure state can be seen in two level in the present proposed approach. The first secure approach was given in the proposed admin layer, as the declared protocol guaranteed that an authentication criterion, which is public key, can only be distributed by the owner of that public key and not by any other third party. So The authenticity of the public key is given and it can be trusted. The second level of the security is given in the nature of proposed multi criteria approach.

While Public key encryption is still a trusted approach, the second criterion provides more security in a way that only those involved parties in a communication have the required information. A relation/session based information will be generated at the end of each authentication process to be used in next authentication process. Using the described Diffie Hellman protocol in Sect. 4.3 to generate a one time usable encryption key, is very important since the used key is not stored anywhere and can not be compromised over the time. In Opposite to the Public key, it will be recreated each time after the usage. Even after the production of a new shared credential using that secret key, the credential is not exchanged over the net, but only the proof, that the partner in possession of that credential is, will be delivered. So no third party is able to reproduce a fake authentication proof in between. This way the proposed multi criteria authentication approach is 100% secure and the third condition is also guarantied.

5 Conclusion

In this paper a solution has been proposed to support a multi criteria authentication approach in P2P networking environment without any centralized instance involved. To achieve this goal, a dedicated multidimensional administration graph has been designed as infrastructure to maintain peers public key among other information. This graph stores also community membership of peers and resources they want to publish in the community. The proposed infrastructure ensures the accessibility of each peers Information, including resources they provide and their authentication credentials, in the network for every searcher visible. That means the infrastructure provides a global view in the network in a fully distributed manner. Based on abilities enabled in this infrastructure, a secure multi criteria authentication has been supported to ensure a secure and trustable authentication process.

References

1. Novaković, N., Latinović, M.: PKI systems, directives, standards and national legislation. AKTUELNOSTI **2**(32) (2017)
2. Rivest, R.L., Shamir, A., Tauman, Y.: How to share a secret. Commun. ACM **22**, 612–613 (1979)
3. Kurihara, J., Kiyomoto, S., Fukushima, K., Tanaka, T.: A new (k,n)-threshold secret sharing scheme and its extension. In: Wu, T.-C., Lei, C.-L., Rijmen, V., Lee, D.-T. (eds.) ISC 2008. LNCS, vol. 5222, pp. 455–470. Springer, Heidelberg (2008). https://doi.org/10.1007/978-3-540-85886-7_31
4. Shamir, A.: Identity-based cryptosystems and signature schemes. In: Blakley, G.R., Chaum, D. (eds.) CRYPTO 1984. LNCS, vol. 196, pp. 47–53. Springer, Heidelberg (1985). https://doi.org/10.1007/3-540-39568-7_5
5. Boneh, D., Franklin, M.: Identity-based encryption from the weil pairing. In: Kilian, J. (ed.) CRYPTO 2001. LNCS, vol. 2139, pp. 213–229. Springer, Heidelberg (2001). https://doi.org/10.1007/3-540-44647-8_13
6. Jagadale, N.N., Parvat, T.J.: A secured key issuing protocol for peer-to-peer network. In: 2014 IEEE Global Conference on Wireless Computing Networking (GCWCN), pp. 213–218 (2014)
7. Vanrenen, G., Smith, S.: Distributing security-mediated PKI. In: Katsikas, S.K., Gritzalis, S., López, J. (eds.) EuroPKI 2004. LNCS, vol. 3093, pp. 218–231. Springer, Heidelberg (2004). https://doi.org/10.1007/978-3-540-25980-0_18
8. Schwoon, S., Wang, H., Jha, S., Reps, T.: Distributed certificate-chain discovery in SPKI/SDSI. In: Proceedings of 15th IEEE Computer Security Foundations Workshop, pp. 129–144 (2002)
9. Gokhale, S., Dasgupta, P.: Distributed authentication for peer-to-peer networks. In: Proceedings of 2003 Symposium on Applications and the Internet Workshops, pp. 347–353. IEEE (2003)

10. Graffi, K., Mukherjee, P., Menges, B., Hartung, D., Kovacevic, A., Steinmetz, R.: Practical security in P2P-based social networks. In: IEEE Society: The 34th Annual IEEE Conference on Local Computer Networks (LCN), pp. 269–272 (2009)
11. Takeda, A., Chakraborty, D., Kitagata, G., Hashimoto, K., Shiratori, N.: Proposal and performance evaluation of hash-based authentication for P2P network. J. Inf. Process. Inf. Process. Soc. Japan 59–71 (2009)
12. Tsang, P.P., Smith, S.W.: PPAA: peer-to-peer anonymous authentication. In: Bellovin, S.M., Gennaro, R., Keromytis, A., Yung, M. (eds.) ACNS 2008. LNCS, vol. 5037, pp. 55–74. Springer, Heidelberg (2008). https://doi.org/10.1007/978-3-540-68914-0_4
13. Alfarez, A.-R.: The PGP trust model. EDI-Forum J. Electron. Commer. **10**(3), 27–31 (1997)
14. Naranjo, J., Cores, F., Casado, L.G., Guirado, F.: Fully distributed authentication with locality exploitation for the CoDiP2P peer-to-peer computing platform. J. Supercomput. **65**(3), 1037–1049 (2013)
15. Capkun, S., Buttyäin, L., Hubaux, J.-P.: Self-organized public-key management for mobile ad hoc networks. IEEE Trans. Mob. Comput. **2**, 52–64 (2003)
16. Nakamoto, S.: Bitcoin: a peer-to-peer electronic cash system. Decentralized Business Review (2008)
17. Moradi, J., Shahinzadeh, H., Nafisi, H., Gharehpetian, G.B., Saneh, M.: Blockchain, a sustainable solution for cybersecurity using cryptocurrency for financial transactions in smart grids. In: 2019 24th Electrical Power Distribution Conference (EPDC), pp. 47–53 (2019)
18. Hammi, M.T., Bellot, P., Serhcouhni, A.: BCTrust: a decentralized authentication blockchain-based mechanism. In: 2018 IEEE Wireless Communications and Networking Conference (WCNC), pp. 1–6. IEEE (2018)
19. Liu, B., et al.: A new group-to-group authentication scheme based on PUFs and blockchain. In: 2019 IEEE 4th International Conference on Signal and Image Processing (ICSIP), pp. 279–283 (2019)
20. Mukhandi, M., Damião, F., Granjal, J.: Blockchain-based device identity management with consensus authentication for IoT devices. In: 2022 IEEE 19th Annual Consumer Communications & Networking Conference (CCNC), pp. 433–436. IEEE (2022)
21. Fanti, G., Kairouz, P., Oh, S., Viswanath, P.: Spy vs. spy: rumor source obfuscation. In: Proceedings of the 2015 ACM SIGMETRICS International Conference on Measurement and Modeling of Computer Systems, pp. 271–284. ACM (2015)
22. Chaum, D.: The dining cryptographers problem: unconditional sender and recipient untraceability. J. Cryptol. **1**, 65–75 (1988)
23. Diffie, W., Hellman, M.: New directions in cryptography. IEEE Trans. Inf. Theory **22**(6), 644–654 (1976)
24. Fanti, G., Viswanath, P.: Algorithmic advances in anonymous communication over networks. In: 2016 Annual Conference on Information Science and Systems (CISS), pp. 133–138. IEEE (2016)
25. Feigenbaum, J., Ford, B.: Seeking anonymity in an internet panopticon. Commun. ACM **58**(10), 58–69 (2015)
26. Goldwasser, S., Micali, S., Rackoff, C.: The knowledge complexity of interactive proof systems. SIAM J. Comput. **18**(1), 186–208 (1989)
27. Kurmi, J., Sodhi, A.: A survey of zero-knowledge proof for authentication. Int. J. Adv. Res. Comput. Sci. Softw. Eng. **5**(1), 494–501 (2015)

28. Pang, X., Wang, C., Zhang, Y.: A new P2P identity authentication method based on zero-knowledge under hybrid P2P network. TELKOMNIKA Indonesian J. Electr. Eng. **11**(10), 6187–6192 (2013)
29. Martín-Fernández, F., Caballero-Gil, P., Caballero-Gil, C.: Authentication based on non-interactive zero-knowledge proofs for the internet of things. Sensors **16**(1), 75 (2016)

Reduction of Overhead for Protocols with Remote Memory Properties

Dimitri Samorukov$^{(\boxtimes)}$

32457 Porta Westfalica, Germany
dimitrisamorukov@yahoo.de

Abstract. Master-slave communication via protocols with remote-memory properties can contain significant overheads. An example of these protocols is MODBUS. Here is shown that user data-oriented optimisation of access to the slave can be substantially optimised if the data to be transmitted is grouped into fewer requests. However, the optimal aggregation of requests can only be achieved with considerable computational effort. So this work relies on a stochastic solution calculation based on the Great Deluge Algorithm. Compared to a random request aggregation, the optimised stochastic request aggregation can result in up to 35% less transmitted data.

Keywords: MODBUS · industrial communication · real time · network · optimization

1 Introduction

Automation systems are used in many areas from building automation to the control of production processes. Generally, these consist of a central control unit with several subsystems (called slaves here) connected via different communication networks.

The control unit often corresponds to a PLC (according to IEC61131), which also acts as a master for the respective communication network. The PLC executes its program cyclically. The execution is subject to hard real-time requirements, i.e. the cycle duration must not exceed an upper time limit. Before the cycle, the current values of all slaves are collected via the respective communication network. The new values are written to the slaves after the cycle.

As a communication network, the Ethernet-based methods (PROFINET [1], EtherCAT [3], MODBUS TCP/UDP [4]) are increasingly replacing the older fieldbus systems (PROFIBUS, CANopen, IO-Link [2], MODBUS RTU). To meet the hard real-time requirements, changes must be made to the data link layer (ISO OSI Layer 2) of the standard Ethernet protocol. The communication for each participant is planned in advance. However, parallel TCP/IP communication (such as MODBUS TCP/UDP) is still possible in these networks. It is of great interest for the operator to use the available bandwidth as efficiently as possible.

A slave can be a simple sensor/actuator up to a complex production machine. Its representation in the respective communication network (i.e. which data it provides to the master for reading or writing) is determined by the respective specification. It is referred to as the slave model. The slave model differentiates the data

H. Unger and M. Schaible (Eds.): Real-Time 2022, LNNS 674, pp. 213–231, 2023.
https://doi.org/10.1007/978-3-031-32700-1_19

according to real-time requirements (cyclic, acyclic). In addition, the type of organisation of the acyclic data can be distinguished (hierarchical, key-value pairs, flat). Cyclic data of the slave are also referred to as process data, acyclic data form the configuration data. Whereby the process data are subject to the respective real-time requirements (must be exchanged cyclically with the PLC), the configuration data are only exchanged occasionally (e.g. in the initialisation phase) with the master. Many communication networks (such as PROFINET [1], EtherCAT [3]) determine in the initialisation phase (connection establishment) which data are exchanged as process data. Individual values of the process data are referred to as data points, of the configuration data as parameters.

The organisation of the acyclic data of a slave is determined by the specification of the respective communication network. PROFINET [1] and IO-Link [2], for example, require a hierarchical structure of the data of a slave so that each parameter can be reached by means of a module/submodule/index/subindex path within a slave. Others, such as EtherCAT [3] or CANOpen organise all parameters within a key-value pair directory. Here, each parameter can be addressed by means of a key. The simplest organisation of the parameters is found in MODBUS [4], where all parameters are arranged one after the other in a flat list[1]. The user can access the parameters (or their elements) either register by register (16 bit) or bit by bit. Which parameter (data type, meaning) is behind an address can be found either in the specification (well known parameter) or in the respective manufacturer's description of the slave (device description).

MODBUS [4] is a classic master/slave protocol where the master cyclically reads or writes the required data points/parameters from the slave. The requests contain the start address and the number of values to be read/written. There is no specified connection establishment or termination. The slave model is constructed in such a way that there is no clear distinction between parameters and data points in the flat memory available. Behind each address in the memory there is either a register value or a bit value (coil).

MODBUS supports different types of requests that can read/write different numbers of values. Each request transmits additional administrative data (IP header, TCP header, etc.), called here protocol overhead. One or more data points can be read/written with a request, but values that are not needed will also be transmitted if the requested data points are not directly behind each other, this unused data is called here payload overhead. The problem that is considered here is the following. The user needs n data points from the flat memory of the slave. How must the requests be structured (request type, with start address and length) so that both overheads are minimised but all n required data points are still transmitted? In their overview, Scanzio, Wisniewski and Gaj [10] describe the current state of communication protocols in industrial automation. In particular, it becomes clear how heterogeneous communication

---

[1] Organisation by key-value pairs is also allowed, but is not considered further in this work.

protocols are (and are likely to remain) and how important it is to use existing communication resources effectively.

This work has the following structure. First, in Sect. 2 the system under consideration is further formalised. Based on this formalisation, the problem is described in Sect. 3. For the described problem, two solutions are now offered (Sect. 5), an ideal solution, which, however, is practically not realisable, and then an approximated version. The approximated solution is then verified in Sect. 6. The work is concluded with Sect. 7.

2 System

This section introduces the formal model of the system. Firstly, however, an informal description clarifies what is meant here by protocol- and payload overheads and what the flat-list model of a slave looks like.

A MODBUS [4] request consists of two transmitted packets, one packet to the slave and the response from the slave. A packet always consists of a MODBUS header and payload data. The header data is considered here as the protocol overhead. These appear in the respective request and the response. Table 1 gives an overview of the respective overhead for the request, response, sum of both and maximum payload data in bytes for the common request types (FunctionCodes, FC's). This table does not contain further protocol overhead which is additionally caused by e.g. TCP/UDP layer, IP layer and the Ethernet layer. These typically amount to 66 additional bytes (26 bytes Ethernet-II IEEE 802.3, 20 bytes IP and 20 bytes TCP/UDP) per direction, which makes 126 bytes per request and the response. A complete MODBUS request is also referred to in the following as a *request-response cycle*.

Table 1. MODBUS function codes with request and response protocol overheads in bytes

FunctionCode (FC)	Req.	Resp.	Sum	Payload
Read coils(1)	5	2	7	2000
Read discrete inputs(2)	5	2	7	2000
Read holding registers(3)	5	2	7	2000
Read input registers (4)	5	2	7	2000
Write single coil (5)	5	5	10	1
Write single register (6)	3	5	8	16
Write multiple coils (15)	6	5	11	1986
Write multiple registers (16)	6	5	11	1984
Write masked registers(22)	7	7	14	16
Read/write multiple registers(23)	10	2	12	3936

Slave model

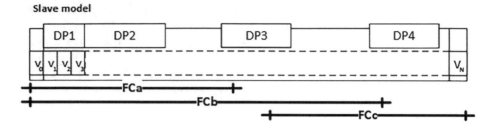

Fig. 1. Example of a slave model in a flat list design

The Fig. 1 shows a example slave model in the flat-list version. Each bit (v) is accessible by at least one FunctionCode (for reading, writing or both). The bits can be grouped into meaningful, application-specific data points (DPx). For example, if a slave represents a temperature sensor, a data point may contain the current measured value. A data point (DPx) always covers n related bits v. There is at least one FunctionCode with which a data point can be accessed (read or write), i.e. all bits of the data point must support this FunctionCode. Several data points can be accessed with one FunctionCode if the max. payload of the corresponding FC is not exceeded and all bits of the data points support the corresponding FCs. In the example, DP1 and DP2 can be accessed with FCa, but not DP2 and DP4 simultaneously.

In the Fig. 1, DP1 and DP2 can be accessed with FCa without gaps. Simultaneous access (with the same response cycle) to DP2 and DP3 is not possible without transmitting the bits in between. These gaps that occur when accessing multiple data points are referred to in the following as the payload overhead.

The configuration procedure of the MODBUS master (e.g. PLC) is therefore as follows. The user takes the model of the slave (with available addresses (v) with their FCs, the existing data points) from a provided device description. He then selects the data points that are of interest to him. With the help of the slave model, corresponding function codes are configured and then accessed cyclically by the master.

2.1   Formal Model

Slave S is defined of a set $S = \{v_1, v_2, ...v_n\}$ of available addressable values. Each *addressable value* is defined by $v = (a, CMD^v)$ with $a \in \mathbb{N}$ as address of the value within the fieldbus image of the slave and $CMD^v = \{cmd^1, cmd^2, ..cmd^N\}$ as set of *commands* that can be used to access (read or write) the value v. For v_i, except the last address v_n, applies $\forall v_i \in S : (\exists! v_p : a(v_p) = a(v_i) + 1)$. Thus the slave has a continuous fieldbus image with unique addresses a. CMD^v is a subset $CMD^v \subseteq CMD$, which holds all known *commands*.

Each *command* is defined by $cmd = (L_{payload}^{max}, L_{overhead}^{prot})$. With $L_{payload}^{max}$ as maximal transmitted payload and $L_{overhead}^{prot}$ as at minimum expected overhead created by a single request/response execution of the *command*, also called protocol overhead. For both values applies $L_{payload}^{max} \in \mathbb{N}$, $L_{overhead}^{prot} \in \mathbb{N}$

Data points for which the addressable values v can be combined are described by: $d_p = (a_{start}, L_{dp}, CMD^{dp})$. a_{start} defines the address of the first addressable value, L_{dp} specifies of amount of subsequent addressable values, including the first one. L_{dp} and a_{start} define a subset of the addressable values from Slave S, $V^{dp} \subseteq S$. This set V^{dp} is used to determine the *command* set CMD^{dp}, with $CMD^{dp} = \{CMD^v(v_x^1) \cap CMD^v(v_x^2)..\cap CMD^v(v_x^N)\}$, where $v_x \in V^{dp}$. It applies $CMD^{dp} \neq \emptyset$ by definition. This means that each element v of the data point d_p can be accessed with all *commands* from set CMD^{dp}.

A *request-response cycle* $c_{rr} = (DP_{rr}, CMD^{rr}, L_c, L_s)$ consists of a data points set $DP_{rr} = \{d_p^1, d_p^2...d_p^N\}$, a set of commands $CMD^{rr} = \{cmd^1, cmd^2, ...cmd^N\}$ that can be used to read the data points. L_s is the start address, defined as $L_s = min(d_p(a_{start}) \in DP_{rr})$. L_c is defined as the number of accessed *addressable values* by the c_{rr}, it is calculated from the set DP_{rr} with $L_c = max(d_p(a_{start}) + d_p(L_{dp}) \in DP_{rr}) - L_s)$

The c_{rr} is *valid* if:

- the set CMD^{rr} is not empty and all commands *cmd* from this set are able to access at least L_c *addressable values* from Slave S
- all available addressable values $v \in S$ with $L_s \leq a \leq (L_s + L_c)$ support all commands from CMD^{rr}
- all *data points* from DP_{rr} do not overlap other data points from DP_{rr}. If $A^i = \{a^1, a^2..a^n\}$ is the set of accessed address by a data point d_p^i, then it applies $\{A^1 \cap A^2.. \cap A^N\} = \emptyset$ for all *data points* from DP_{rr}

The *payload overhead* P_{ov} of a request-response cycle c_{rr} is defined as difference between all transmitted bytes L_c and the requested bytes as the sum $\sum L_{dp}$ of all *data points* V^{dp}, it applies:

$$P_{ov} = L_c - \sum L_{dp}. \tag{1}$$

The *protocol overhead* P_{pr} of a request-response cycle c_{rr} is defined as $min\{L_{overhead}^{prot}(cmd^1), L_{overhead}^{prot}(cmd^2), ..\}$ over all $cmd \in CMD^{rr}$.

3 Problem Definition

The user now wants to access the slave S as efficiently as possible. He has a slave description S and he knows which data points D_p^{in} he wants to access. What he does not know, however, is how the requests to the slave must be put together in order to access the required data points with the smallest possible overhead (generated network load).

This can also be expressed in a formal way as follows. User defines a set of *data points* D_p^{in} for a slave S. The problem now is to generate a set $C_{rr}^{res} = \{c_{rr}^1, c_{rr}^2, ..c_{rr}^N\}$ of *response-request cycles* in such a way that each requested *data point* $d_p \in D_p^{in}$ is contained in exactly one *response-request cycle* c^{rr} and the amount of *payload overhead* P_{ov} and P_{pr} *protocol overhead* over all *response-request cycles* c^{rr} is minimized. It applies

$$\sum_{c^r \in C^{res}} P_{ov}(c^r) + P_{pr}(c^r) = min \qquad (2)$$

4   Related Work

The automation systems considered here are primarily concerned with meeting the hard real-time requirements of the PLC. To do this, the data that is exchanged via the network must be known. Then it is possible to intervene on two levels. First, the point-to-point data flow (master-slave) must be planned (point-to-point planning), which can then be used to plan the individual packet transmissions in the entire network (network planning).

Network planning expects information about all packets, their cycle time, source and sink as input. From this, the paths of the packets in the physical network can be planned using network calculus methods [5,6], so that the packets reach their destinations in time. Regardless of the methods mentioned later, the appropriate communication protocol must already be used in the planning. The individual packet and summation frame approaches are the main distinguishing features of the two communication protocols EtherCAT and PROFINET (and other like POWERLINK, SERCOS III etc.). Depending on the specific application or network topology, the decision in favour of one or the other method can already significantly reduce the transmission volumes on Ethernet (see [13]).

Here, point-to-point planning is considered first and foremost. This can be optimised in three ways. a) The slave can perform preprocessing of the data and thus minimise the transmitted packet length. b) Connection-oriented methods can be used so that the sizes of the individual exchanged packets can be minimised after the connection is established. c) The master makes its requests in such a way that the slave has to transmit as few data as possible.

Examples of data pre-processing in the slave can be found in PROFINET and IO-Link. IO-Link [2] structures the parameters of the acyclic data of a slave hierarchically, with index/SubIndex addresses. By means of SubIndex, each parameter can be addressed individually. However, if the SubIndex 0 of an index is accessed, all parameters of the respective index are addressed. Thus, the protocol overhead can be reduced considerably if all parameters of a index have to be accessed. This applies to both the reading of services (AL_Read) and the writing of parameters (AL_Write). Like IO-Link, PROFINET [1] relies on a hierarchical structure of the acyclic parameters, addressing takes place by means of Slot/SubSlot/Index. Behind each of these addresses are several parameters in a flat list (Record Data object). The object can contain many parameters. This can cause a lot of payload overhead when only a few parameters are needed. Thus, the concept of the selector was introduced. By means of an additional addressing level Slot/SubSlot/Index/Selector, the slave can offer a reduced set of parameters of the record data object. In this way, small parameters (a few bits) can be accessed more effectively with less payload overhead (see service Read Query). Unfortunately, the selector concept is not dynamic. It cannot be adapted to the respective requirements at runtime. Parameters behind a selector are defined by the manufacturer of the slave at the time of production.

In the transmission of cyclic data, connection-oriented procedures can be assumed. For this purpose, the master and slave establish a connection; in the initialisation phase, it is determined which data points are exchanged in which direction. This information does not have to be retransmitted later and the packet size (and thus the protocol overhead) can be reduced. This is the case with both PROFINET [1] and EtherCat [3]. With PROFINET, both the slave and the master can send a packet (individual packet), and with EtherCat, only one packet (summation frame approach) is sent from the master to the slaves, which then enter or extract their data at the respective position in the packet.

The solution chosen in this work falls into the third category, in which the master makes the request in such a way that the slave has to transmit as little as possible. The focus here is restricted to MODBUS slaves. From the historical perspective, accessing them is the same as accessing a remote memory, but the bandwidth is significantly limited. With this view, it is very effective to reduce the use of bandwidth by avoiding the requests completely. For this purpose, the well-known caching mechanisms (e.g. [9]) exist so that the master only reads the data that it has not read for a long time. Unfortunately, these methods are only conditionally suitable for automation systems, because other masters or the slave itself can change its data points or parameters.

The solution provided here is inspired by the reinforcement learning method, where the algorithm, which is executed cyclically and must choose between available operations, must find an order of operations, so that the long-term reward is maximised (see [11,12], multi-armed bandit problem). Each operation brings a direct reward. However, the order of operations chosen has an impact on the total reward yield. Applied to this problem here, it means that the merging of requests is repeated several times, and the order of merge operations of requests is retained at the end that leads to the greatest reward, i.e. to the smallest overhead here. The optimal order of merge operations must be found in such a way as to reduce the final communication volumes (protocol and payload overheads).

5   Solution

The idea behind the solution approach is based on the fact that first a set C_{rr}^0 of *response-request cycles* is generated from the initial data point set D_p^{in}. C_{rr}^0 contains a single *request response cycle* for every *data point*.

The first generation C_{rr}^0, however, has a very high *protocol overhead* P_{pr} because there are many *response-request cycles*. Therefore, their number must be reduced and the available payload $L_{payload}^{max}$ must be better utilised. For this reason, several merges of the *response-request cycles* are executed. Each merging chooses two c_{rr} *response-request cycles* from current generation C_{rr}^x and generates a new, reduced by one generation C_{rr}^{x+1} of *response-request cycles*.

This takes place as long as further valid *response-request cycle* can be generated after merging and the amount can be reduced (new generation C_{rr}^{x+1} can be created). See therefore the Fig. 2.

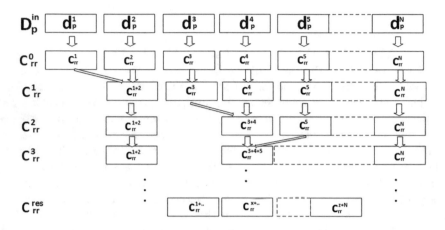

Fig. 2. Reduction of response-request cycle by combining them

The function merge ($c_{rr}^z = merge(c_{rr}^x, c_{rr}^y)$) expects two *response-request cycles* (c_{rr}^x, c_{rr}^y) as input. The output is a new, valid *response-request cycle* c_{rr}^z that contains all data points from c_{rr}^x and c_{rr}^y, $c_{rr}^z = (DP_{rr}^z, CMD_{rr}^x, L_c^x)$ with $DP_{rr}^z = DP_{rr}^x \cap DP_{rr}^Y$ and CMD_{rr}^z, L_c^z are calculated as described above. If the creation of new valid c_{rr}^z is not possible, c_{rr}^z will be $\emptyset$. It should be noted that $merge(c_{rr}^x, c_{rr}^y) = merge(c_{rr}^y, c_{rr}^x)$ applies.

In order to meet the requirements of Eq. 2, the *response-request cycle* selected for merge must be chosen so as to reduce the payload overhead P_{ov}.

5.1   Exhaustive Search

The most obvious way to find the optimal solution, or the optimal couple c_{rr}^x, c_{rr}^y, is to explore all possibilities and then choose the one with the lowest overhead. See Fig. 3. It is started here with the C_{rr}^0 with 4 cycles. It is assumed that the merging is always possible. For all possibilities to combine the 4 *response-request cycles* with each other, a set of next lower level generations $CL_{rr}^1 = \{C^11_{rr}, C^12_{rr}, ..\}$ is generated. This results in 6 combinations to form the next generation. The same is done again with the now created next generations C_{rr}^1. This is repeated until the last possible generation C_{rr}^3 is created. From this generations overheads P_{ov} and P_{pr} are calculated. The result of this is the pair $\{c_{rr}^x, c_{rr}^y\} \in C_{rr}^0$ whose combination leads to the minimal P_{ov} and P_{pr}. In the real examples not every combination will be allowed. So there are fewer possibilities to try out.

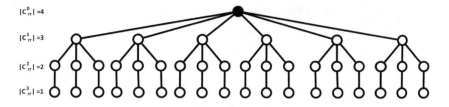

$|c_{rr}^0|=4$

$|c_{rr}^1|=3$

$|c_{rr}^2|=2$

$|c_{rr}^3|=1$

Fig. 3. Possible merges with 4 initial *response-request cycles*

There are $N_{mrg}(m) = \binom{m}{2}$ possibilities to create generations of the next shorter length (next lower level generation set CL_{rr}^1) from the initial generation C_{rr}^0 with $m = |C_{rr}^0|$. So there are also $\binom{m}{2}$ different merge operations possible. The Fig. 4 gives an example with $m = |C_{rr}^0| = 3$, here are 3 different merge operations possible to create the set $CL_{rr}^1 = \{C_{rr}^{11}, C_{rr}^{12}, C_{rr}^{13}\}$ and 6 different merge operations to create $CL_{rr}^2 = \{C_{rr}^{21}, C_{rr}^{22}, C_{rr}^{23}\}$

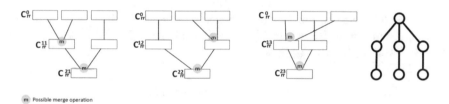

m Possible merge operation

Fig. 4. Possible merges tree: possible merge operations from the same root generation C_{rr}^0 with 3 *response-request cycles*

The number of all maximal possible merges (ways) to create the deepest possible lower generation set CL_{rr}^d, which consists only of such generations C_{rr}^{dp} with exactly one *response-request cycle* c_{rr} is given by:

$$N_{mrg}^{max}(m) = \binom{m}{2} + \binom{m}{2} \cdot \binom{m-1}{2} + .. \binom{m}{2} \cdot \binom{m-1}{2} \cdot ... \binom{2}{2} \tag{3}$$

$$N_{mrg}^{max}(m) = \sum_{k=2}^{m} \prod_{n=k}^{m} \binom{n}{2} \tag{4}$$

with $m = |C_{rr}^0|$ as the number of *response-request cycles* in the initial generation. These generations C_{rr}^{dp} thus also form the optimal solution sought. Equation 4 is only defined for values greater than 2. The value of N_{mrg}^{max} increases very fast, Table 2 presents some of the values.

The problem of the computational effort of the "exhaustive search" can be reduced by not examining all future possibilities, but only the generations C_{rr} after the next N merges. Thus all merge possibilities starting from the current generation C_{rr}^0 up to and including generations after N - merges are examined

Table 2. Numerical values of Eq. 4

m	N_{mrg}^{max}
3	6
4	42
5	430
10	6159299985

and the pair c_{rr}^x, c_{rr}^y from the current generation C_{rr}^0 is selected, which leads to smallest overhead P_{ov} and P_{pr}.

The number of possible merges to create next lower level generations set $CL_{rr}^1, CL_{rr}^2, ..CL_{rr}^z$ is given by:

$$N_{mrg}^{nx}(m, z) = \sum_{k=z+1}^{m} \prod_{n=k}^{m} \binom{n}{2} \tag{5}$$

, with $m = |C_{rr}^0|$ as the number of *response-request cycles* in the current generation C_{rr}^0, and $z = |C_{rr}^x|$ as size of smallest reachable C_{rr}^x . This is only defined for $m \geq 2$ and $z \geq 1$ with $m \geq z$.

Table 3. Numerical values of Eq. 5

m	z	$N_{mrg}^{nx}(m, z)$
5	2	250
5	3	70
10	2	3587387985
10	3	1015475985
10	5	15287985

Even if this approach is able to find the most optimal solution, i.e. the best choice $\{c_{rr}^x, c_{rr}^y\} \in C_{rr}^0$ at any time, its practical application fails due to the required computing capacity (see Table 2). A significant calculation effort is still necessary (see Table 3), even if not all possibilities are tried out, but the search only looks only a few merges into the future.

5.2   Stochastic Approach

$N_{mrg}^{max}(m)$ gives the number of merges to reach all possible c_{rr} starting from generation C_{rr}^0. The minimal number of merges to reach the optimal solution is $N_{mrg}^{min} = |C_{rr}^0| - 1$, after N_{mrg}^{min} merges there will remain exactly one *response-request cycle* c_{rr} which is the optimal solution, if all merges were valid.

The question is how to find two optimal $\{c_{rr}^x, c_{rr}^y\}$ *response-request cycles* from the initial generation C_{rr}^0 to reach the generation C_{rr}^e with minimal P_{ov} and P_{pr} without the need for exhaustive search.

This can be satisfied by the Great Deluge algorithms (GD) [8], which is applied here for the stochastic approach. In this context, the following terms need to be clarified:

- solution SL and it's fitness O
- creation of a solution SL, initial solution SL_i and next solution SL_x
- best solution SL_b and it's fitness O_b
- fitness tolerance threshold FTH
- idle steps and according limit N_{idle}^{GD}.

The GD algorithm takes an initial solution SL_i given and generates a next solution SL_x from it. If the fitness O this solution is under the tolerance threshold FTH, this solution is retained and the process is now repeated with the solution SL_x. If a new best value for fitness O was reached with the created solution, this solution is also retained as the current best solution SL_b. The tolerance threshold FTH is increased with each best solution. If FTH has reached the fitness value of the best solution, each newly generated solution must have a fitness above the fitness of SL_b, if it is to be retained. The algorithm is terminated if no new best solution is found after N_{idle}^{GD} new generated solutions (GD idle steps).

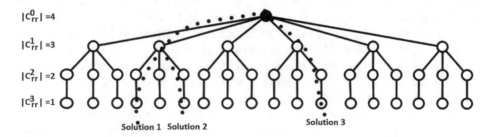

Fig. 5. Solution generation on possible merges tree

A solution $SL = \{sl_1, sl_2..sl_N\}$ is a ordered sequence of merge operations

$$sl = \{\{\{c_{rr}^x, c_{rr}^y\} | c_{rr}^x \in C_{rr}^0, c_{rr}^y \in C_{rr}^0\}, \{\{c_{rr}^x, c_{rr}^y\} | c_{rr}^x \in C_{rr}^1, c_{rr}^y \in C_{rr}^1\}, ...\} \quad (6)$$

starting from the initial generation C_{rr}^0. Thus a solution SL defines a path over all possible generations C_{rr} starting with the initial generation C_{rr}^0, see therefore the Fig. 5. In this figure are three possible solutions drown for a initial generation with four merge-able *request-response cycles*.

The result of applying a solution SL to the initial generation C_{rr}^0 is the final generation C_{rr}^f. The fitness O of a solution SL is defined by:

$$O(SL) = \sum_{c_{rr} \in C_{rr}^f} (P_{ov}(c_{rr}) + P_{pr}(c_{rr})) \tag{7}$$

A metric (called *distance*) was introduced between two data points, see Sect. 5.2. The more advantageous is it to combine two data points, the smaller is the distance. However, it does not take into account the long-term effects, only the improvement of the current generation is considered. This distance exists only for *request-response cycles* that can be merged (their merge leads to a valid new c_{rr}). Effects on the next and following generations are not considered. This distance is used to calculate the initial and the next solutions.

For the generation of the initial solution SL following approach is used. From the initial generation C_{rr}^0 two *response-request cycles* $\{c_{rr}^1, c_{rr}^2\} \in C_{rr}^0$ are selected. The first c_{rr}^1 is the one with smallest address v. The second *response-request cycle* is the one with smallest *distance* (if it exists) to the first c_{rr}^1 *response-request cycle*. Both cycles are the first entry in $\{c_{rr}^1, c_{rr}^2\} \in SL_i$. They are merged and so new generation C_{rr}^1 is created. This steps are repeated as long as further two *request-response cycles* $\{c_{rr}^1, c_{rr}^2\} \in C_{rr}^x$ can be found. Until here no stochastic approaches are used.

The initial solution SL_i becomes the current solution SL_x. From this current solution SL_x, next solution SL_{x+1} is generated by slightly, randomly modifying the current solution (consider Fig. 5). The modifying of the current solutions means that on random step (random $sl \in SL$ and according generation C_{rr}) other than original $\{c_{rr}^x, c_{rr}^y\}$ *response-request cycles* are selected for merge. This leads to creation of other C_{rr} than in the original solution SL_x. Selection of entry $sl \in SL_i$ is done by uniform distribution. SL_{x+1} is initialized with all entries from SL_x until sl. Then the according generation C_{rr}^s is created by executing the merge operations starting from C_{rr}^0 until the merge represented by the selected entry sl. Now two *response-request cycles* from C_{rr}^s are randomly selected. The first $c_{rr}^1 \in C_{rr}^s$ is selected using uniform distribution. The selection of second $c_{rr}^2 \in C_{rr}^s$ is done also randomly but a distance-based distribution is used. Therefor the *distance* from c_{rr}^1 to every other $c_{rr} \in C_{rr}^s$ is calculated. This distances are used to calculate the cumulative distribution function (CDF). With inverse transform sampling method [7] and a random number selected from uniform distribution the second entry $c_{rr}^2 \in C_{rr}^s$ is selected. The new entry $sl^x = \{c_{rr}^1, c_{rr}^2\}$ is appended to SL_{x+1} and the next generation C_{rr}^{s+1} is created. The creation of next generation and the selection of a pair $\{c_{rr}^x, c_{rr}^y\}$ is repeated again, as long as a pair can be found. Otherwise the creation of SL_{x+1} is done.

For newly found solution SL_{x+1} the fitness O (see formula 7) is calculated. If O is over the current threshold FTH value, SL_{x+1} will be discarded and other SL_{x+1} from SL_x will be calculated. If O is better then current O_b, then N_{best} is increased, SL_b is set to SL_{x+1}, O_b is set to O and FTH is set to new value by the following formula:

$$fct(N_{best}) = \begin{cases} \frac{1-OF^{max}}{N_{best}^{max}} \cdot N_{best} + OF^{max} & : \quad 0 \leq N_{best} \leq N_{best}^{max} \\ 1 & : \quad \text{else} \end{cases} \tag{8}$$

$$FTH = fct(N_{best}) \cdot O \tag{9}$$

where OF^{max} is the maximum allowed fct-value, N_{best}^{max} is the number of found best solutions after them the $fct(N_{best}^{max}) == 1$ applies. OF^{max} and N_{best}^{max} are parameter of the solution search. Thus, the FTH falls linear with each new SL_b found until it equals the fitness O of the current SL_b after N_{best}^{max} found best values.

Distance Between Request-Response Cycles. A distance between two cycles c_{rr}^x, c_{rr}^y is defined only for pairs of *response-request cycles* that can be merged (result is a valid c_{rr}):

$$dst(c_{rr}^x, c_{rr}^y) = \frac{P_{ov}^{xy} + P_{pr}^{xy}}{|CMD_{xy}^{rr}|} \tag{10}$$

By this definition the distance $dst(c_{rr}^x, c_{rr}^y)$ between two c_{rr} *response-request cycles* decreases with the number of different possible commands $cmd \in CMD_{xy}^{rr}$ and increases with the expected *payload- and protocol overhead* $P_{ov}^{xy} + P_{pr}^{xy}$ after merging of the both *response-request cycles*.

6 Verification

6.1 Hypotheses

The following hypothesis are to be verified here. It is to be shown that

I) the stochastic approach (based on the Great Deluge algorithm) is better in average than the approach that randomly selects the $\{c_{rr}^x, c_{rr}^y\}$ cycles from the current generation C_{rr} for the next merge operation (*random solution search*) and

II) the stochastic approach finds the better solutions the more different commands support the addresses of the slave, compared to the *random solution search*.

In detail, this means that the fitness values (see formula 7) of the two solutions (initial and final) of the stochastic approach are compared with the fitness value of the *random solution search*.

6.2   Setup

The verification takes place simulatively. The structure of the simulation is described below. Firstly, however, the *random solution search* is defined.

Random solution search takes as input an initial generation C_{rr}^0 and generates a list of possible merge operations $SL = \{sl_1, sl_2..sl_N\}$ with
$$sl = \{\{\{c_{rr}^x, c_{rr}^y\} | c_{rr}^x \in C_{rr}^0, c_{rr}^y \in C_{rr}^0\}, \{\{c_{rr}^x, c_{rr}^y\} | c_{rr}^x \in C_{rr}^1, c_{rr}^y \in C_{rr}^1\}, ...\}$$
starting from the initial generation C_{rr}^0 (see also the solution definition in Sect. 5.2). In difference to stochastic approach, the selection of $sl = \{c_{rr}^x, c_{rr}^y\}$ from the current generation C_{rr} is executed randomly. The first c_{rr}^x is chosen randomly from all $c_{rr} \in C_{rr}^x$. The second $c_{rr}^y \in C_{rr}^x$ is chosen randomly from all remaining response-request cycles that can be merged into a valid c_{rr} (equal distribution). If no matching $c_{rr}^y \in C_{rr}^x$ is found for the selected c_{rr}^x, a next $c_{rr}^x \in C_{rr}^x$ is selected. The process is repeated as long as no valid pair $\{c_{rr}^x, c_{rr}^y\}$ is found in the current C_{rr}^x. If a valid pair is found. The next generation C_{rr}^{x+1} is generated by merging of $\{c_{rr}^x, c_{rr}^y\}$. The search for a solution is completed if no merge-able pair was found in the current C_{rr}^x.

Each simulation consists of N_{steps}^{sim} individual steps. In each step, a new slave S is built (see below). In the slave, N_{dp} data points d_p with a maximum length of L_{dp}^{max} are now generated. Exactly one c_{rr} cycle is generated from each data point. This initial generation C_{rr}^0 is now merged as far as possible using *random solution search* and the stochastic method (GD algorithm). For each new simulation step, the two fitness values (initial O_{init} and final O_b) of the GD solution and the fitness O_{rnd} of the *random solution search* are recorded. The two GD fitness values are normalised to the value O_{rnd} of the *random solution search* in each simulation step, with $O_{init}^{nrm} = O_{init}/O_{rnd}$ and $O_b^{nrm} = O_b/O_{rnd}$. The respective mean values ($\overline{O_{rnd}}, \overline{O_{init}}, \overline{O_b}, \overline{O_{init}^{nrm}}, \overline{O_b^{nrm}}$) are calculated from all recorded values. In addition, the number of simulation steps in which the a) initial solution was better than the random solution (r_{init}^f) and b) the final solution was better than the random solution (r_b^f) are counted. These two values are called the reliability ratio and are given as a percentage of all simulation steps in the respective simulation. The values $r_b^f, r_{init}^f, \overline{O_{rnd}}, \overline{O_{init}}, \overline{O_b}, \overline{O_{init}^{nrm}}, \overline{O_b^{nrm}}$, form the result of a simulation.

The slave build process takes as input number of available slave addresses $|S|$, set of all known commands CMD, number of commands to use (N_{cmd}), number of data points to create N_{dp}, maximal length L_{dp}^{max} of the created data points. First the set $S = \{v_1, v_2..n\}$ with $|S|$ addressable values is created. Then CMD^s is created, therefore N_{cmd} random commands from CMD are chosen. For each $cmd \in CMD^s$ and random start $v^{start} \in S$ and random end $v^{end} \in S$ are selected, and all values between v^{start} and v^{end} the cmd is assigned to the set CMD^v of all v. It is ensured here that $a(v^{start}) \leq K^{min}$ and $a(v^{start}) + K_a^{end} \leq a(v^{end}) \leq a(v^{start}) + K_b^{end}$. This way, each $cmd \in CMD^s$ is assigned to at least a quarter of all addresses with a random start position. Addressable values v without assigned commands are assigned a random command $cmd \in CMD^s$.

Now the data points dp are placed in the slave $S = \{v_1, v_2..n\}$. For this purpose for each new data point a random start address $v \in S$ and a random length L_{dp} (between 1 and L_{dp}^{max}) is chosen. The data point is included if no other data points exist in the selected area and all covered addressable values v have at least one common command. This ensures that each data point can also be read out with at least one command. This is repeated until N_{dp} have been created.

6.3   Simulation Parameter

To verify the hypotheses I and II from Sect. 6.1 three simulations with 1, 6, 12 available commands (N_{cmd}) per slave address v were executed.

The Table 4 gives an overview of all parameters used in the simulations. They were chosen closely to real MODBUS slaves. The parameters K^{start}, K_a^{end}, K_b^{end} ensure that each command takes up a noticeable area in the slave, $|S|$ ensures that the slave has enough free addresses to contain the desired number of data points N_{dp}.

The set CMD of all known commands consists of 12 different commands, with different payload $L_{payload}^{max}$ and overhead $L_{overhead}^{prot}$, see Table 5. These include 4 commands of different payload length, each with a different protocol overhead length (between 1% of the payload, up to 1000%).

Table 4. Parameter used for verification of the results

Parameter	Description	Value
N_{idle}^{GD}	Idle steps limit	20
OF^{max}	max. factor fct for the GD-threshold value FTH	10
N_{best}^{max}	Number of best solutions to reach the $fct = 1$	10
$|S|$	Number of addresses v in slave S	1000
L_{dp}^{max}	Max. length of a data point d_p	10
N_{steps}^{sim}	Number of steps per simulation	10000
N_{cmd}	Number of currently used commands per address v	{1,6,12}
N_{dp}	Number of data points in slave S	50
K^{start}	Maximal address for start of command occurrence	$\frac{|S|}{2}$
K_a^{end}	Minimal offset for end command occurrence	$\frac{|S|}{4}$
K_b^{end}	Maximal offset for end command occurrence	$3 \cdot K_a^{end}$

Table 5. Set of known commands

Command	$L_{payload}^{max}$	$L_{overhead}^{prot}$
cmd^1	10	1
cmd^2	50	1
cmd^3	100	1
cmd^4	10	10
cmd^5	50	10
cmd^6	100	10
cmd^7	10	50
cmd^8	50	50
cmd^9	100	50
cmd^{10}	10	100
cmd^{11}	50	100
cmd^{12}	100	100

6.4   Results

The results of the three simulations are shown in Table 6 and the Figs. 6, 7, 8.
Table 7 shows the reliability results.

Table 6. Fitness values and normalized fitness values of all simulation runs

N_{cmd}	$\overline{O_{rnd}}$	$\overline{O_{init}}$	$\overline{O_{init}^{nrm}}$	$\overline{O_b}$	$\overline{O_b^{nrm}}$
1	1416 ± 1193	1232 ± 1193	0.86 ± 0.15	1215 ± 1200	0.83 ± 0.16
6	1123 ± 376	881 ± 330	0.79 ± 0.12	810 ± 327	0.72 ± 0.11
12	997 ± 229	738 ± 182	0.75 ± 0.11	648 ± 174	0.65 ± 0.10

Table 7. Reached reliability ratio values

N_{cmd}	r_{init}^f	r_b^f
1	93.71%	99,56%
6	96.93%	99,78%
12	98,26%	99,94%

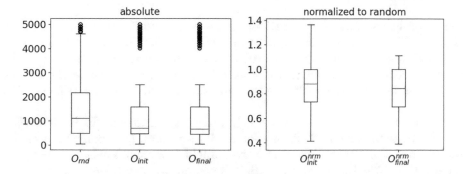

Fig. 6. Simulation results with 1 command

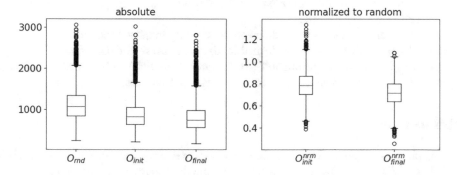

Fig. 7. Simulation results with 6 commands

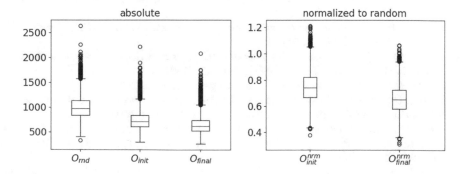

Fig. 8. Simulation results with 12 commands

7 Conclusion and Outlook

The hypotheses I and II from Sect. 6.1 could be proved. By means of reliability r_{init}^f, r_b^f (Table 7) it was shown that in more than 93% of all simulation steps the stochastic, GD-based approach could find a better solution (compared to the *random solution search*), with less *payload and protocol overheads*. This value could even be improved to 99% with more available commands and longer GD search.

On average, the stochastic method improved the fitness of the final solution to 83% (one command), 72% (6 commands) and 65% (12 commands) of the *random solution search*. Thus, the more available commands exist, the better is the solution. The Table 6 and Figs. 6, 7, 8 contain the mentioned results.

Thus it is clear that the described stochastic approach is able to reduce significantly the network load of cyclic communication requests and thus make the utilisation of the available network resources much more economical. The side effect of this solution is that, in addition to reducing the communication effort, the slave and the master are also less burdened and more processor resources are available for other tasks.

A disadvantage here is that the use of stochastic methods yields unpredictable results. In practice, it is often desirable that an input (initial generation C_{rr}^0) always produces the same output (final generation C_{rr}^x). This is not achieved here and should be considered in future work on this basis.

References

1. Application layer services for decentralized periphery and distributed automation (2012). https://www.profibus.com/download/profinet-technology-and-application-system-description. CDV 61158-5-10 ED 3 IEC (E)
2. Io-link interface and system specification (2013). https://io-link.com/share/Downloads/Spec-Interface/IOL-Interface-Spec_10002_V112_Jul13.pdf
3. Ethercat specification (2020). https://www.ethercat.org/
4. Modbus specification (2020). https://modbus.org/docs/Modbus_Application_Protocol_V1_1b.pdf
5. Cruz, R.L.: A calculus for network delay. ii. network analysis. IEEE Trans. Inf. Theory **37**(1), 132–141 (1991)
6. Cruz, R.L., et al.: A calculus for network delay, Part I: network elements in isolation. IEEE Trans. Inf. Theory **37**(1), 114–131 (1991)
7. Devroye, L.: Non-uniform random variate generation (2005)
8. Dueck, G., Scheuer, T., Wallmeier, H.-M.: Toleranzschwelle und sintflut: neue ideen zur optimierung. Spektrum der Wissenschaft **3**(93), 42 (1993)
9. Goodman, J.R.: Using cache memory to reduce processor-memory traffic. In: Proceedings of the 10th Annual International Symposium on Computer Architecture, pp. 124–131 (1983)
10. Scanzio, S., Wisniewski, L., Gaj, P.: Heterogeneous and dependable networks in industry - a survey. Comput. Ind. **125**, 103388 (2021)
11. Silver, D., et al.: Mastering the game of go without human knowledge. Nature **550**(7676), 354–359 (2017)

12. Sutton, R.S., Barto, A.G.: Reinforcement Learning: An Introduction. MIT Press, Cambridge (2018)
13. Wang, K., Li, X., Wang, X.: Analysis of real-time ethernet communication technologies of summation frame and individual frames. In: 2017 IEEE 3rd Information Technology and Mechatronics Engineering Conference (ITOEC), pp. 23–26. IEEE (2017)

Author Index

H. Unger and M. Schaible (Eds.): Real-Time 2022, LNNS 674, pp. 233–234, 2023.
https://doi.org/10.1007/978-3-031-32700-1

Printed in the United States
by Baker & Taylor Publisher Services